FOOD AND BEVERAGE CONSUMPTION AND HEALTH

FERMENTED FOODS

NUTRITION AND ROLE IN HEALTH AND DISEASE

FOOD AND BEVERAGE CONSUMPTION AND HEALTH

Additional books and e-books in this series can be found on Nova's website under the Series tab.

FOOD AND BEVERAGE CONSUMPTION AND HEALTH

FERMENTED FOODS

NUTRITION AND ROLE IN HEALTH AND DISEASE

OLIVER KOVALYOV
EDITOR

nova
science publishers
New York

Library of Congress Cataloging-in-Publication Data

Names: Kovalyov, Oliver, editor.
Title: Fermented foods : nutrition and role in health and disease / Oliver Kovalyov, editor.
Description: New York : Nova Science Publishers, [2020] | Series: Food and beverage consumption and health | Includes bibliographical references and index. |
Identifiers: LCCN 2020041685 (print) | LCCN 2020041686 (ebook) | ISBN 9781536186567 (hardcover) | ISBN 9781536187052 (adobe pdf)
Subjects: LCSH: Functional foods. | Fermented foods--Health aspects. | Fermentation.
Classification: LCC QP144.F85 F47 2020 (print) | LCC QP144.F85 (ebook) | DDC 613.2--dc23
LC record available at https://lccn.loc.gov/2020041685
LC ebook record available at https://lccn.loc.gov/2020041686

Published by Nova Science Publishers, Inc. † New York

CONTENTS

PREFACE

Fermented Foods: Nutrition and Role in Health and Disease provides a comprehensive review of the recently discovered, or bioengineered, vitamin B2, B9 and B12-producing lactic acid bacteria, providing an in depth analysis of the latest biotechnological applications and potentialities, particularly the development of novel bioenriched fermented foods. The authors elucidate the impact of lactic acid fermentation on sulforaphane rich products in an effort to improve our understanding on the role of sulforaphane as a potential medicine in the treatment of various disorders. A proteinaceous compound produced by Leuconostoc lactis RK18 is characterized, isolated from a fermented Cambodian food product. Later, fermented meat products and fermented shrimp products are discussed, particularly focusing on their production, composition, microorganisms, health benefits and health risks.

Chapter 1 - The B group or B-complex vitamins are essential micronutrients possessing physiological effects on a multitude of biological responses, including the immune system. The B-complex vitamins comprise thiamine (B1), riboflavin (B2), niacin (B3), pantothenic acid (B5), pyridoxin (B6), biotin (B7), folate (B9), and cobalamin (B12). Most bacteria, along with plants and fungi, have the ability to biosynthesize *de novo* the vitamins B, whilst humans lack this capacity, hence exogenously acquiring these hydrosoluble molecules from the dietary or microbial sources, namely the endogenous microbiota. Disturbances in the physiological homeostasis of vitamin B elicit a detrimental impact on human health. In this sense, the global upsurge of hypovitaminosis B has been the cause of serious concern, paving the way for the exploitation of suitable alternatives to the conventional chemical-based supplementation. A shifting pattern towards the prevalence of the B-group vitamins deficiencies, previously solely associated with undernutrition and socioeconomic scarcities, has nowadays emerged associated to the renewed vegetarianism-triggered unbalanced diets. In recent years, the development of novel food, pharmaceutical and chemical products have quickly evolved to attend the rising demand for B-group vitamins. Indeed, investigators have concentrated their focus on the exploitation of natural approaches to improve the vitamin B content in food products. The most promising approach is, undeniably, the notable nutraceutical potential of lactic acid bacteria (LAB) as sustainable vitamins biosynthesizers. In the last decades, the development of novel fortified fermented foods, as suitable delivery vehicles to ensure the *in situ* vitamin B bioavailability has been highlighted. These vitamin B bio-enriched products may constitute a feasible prophylactic, or even therapeutic, strategy to counteract hypovitaminosis B. Albeit the significant number of vitamin B-producing LAB identified to date, scarce studies have addressed *in vivo* the impact of the consumption of biofortified functional food products towards the vitamin B deficiency reversion or prevention.

Nonetheless, the scarce bio-based production of LAB relying on intricate metabolic pathways under strict enzymatic regulation poses a prominent obstacle towards the manufacturing scaling-up. Current investigation focuses on genetic bioengineering and integrative multi-omics approaches to generate renewed high-yielding vitamin-producing LAB, and to unveil pivotal enzymes of the metabolic networks underlying the biosynthesis by newly discovered LAB strains. In this chapter the authors present a comprehensive overview of the recently discovered, or bioengineered, vitamin B2, B9 and B12-producing LAB and provide in depth analysis of the latest biotechnological applications and potentialities, particularly, the development of novel bio-enriched fermented foods.

Chapter 2 - Cruciferous vegetables, broccoli, cabbage, cauliflower and vegetable fermented products as well, such as sauerkraut, represent a significant source of sulforaphane (SFN). SFN is a highly bioactive isothiocyanate, derived from its non-bioactive glucosinolate precursor, glucoraphanin (GRP). Upon tissue disruption, β-thioglucosidase enzyme myrosinase or thioglucosidases in intestinal microbiota, convert GRP into SFN and other hydrolysis products. Lactic acid fermentation has been reported to enhance GRP extraction and further SFN bioavailability, through mechanisms yet to be discovered. In addition, several lactic acid bacteria have been reported to possess myrosinase activity, enhancing thus SFN production. SFN metabolism in human and animal cells is performed via mercapturic acid pathway, where SFN is conjugated with glutathione and is subsequently metabolized mainly to its N-acetylcysteine derivative, found in urine and plasma. Its electrophilic nature allows interaction via several signaling cascades, namely Keap1-Nrf2-ARE, MAPK/NF-κB, NLRP3 inflammasome, AMPK, Akt, JAK/STAT, Aβ and tau accumulation, Wnt/β-catenin, Notch and Criptoin prevention or therapy of neurodegenerative, cardiovascular, inflammatory and cancer related diseases. The aim of the present chapter is to elucidate the impact of lactic acid fermentation on SFN rich products and improve the authors' understanding on the role of SFN as potential medicine, in the treatment of various disorders.

Chapter 3 - Safety of fermented foods is mostly ensured by the presence of lactic acid bacteria, which besides having an important role in the enhancement of food shelf-life, also enhance the organoleptic characteristics of the products. One of the major attributes of lactic acid bacteria is their bactericidal activity against spoilage and pathogenic bacteria by the production of proteinaceous substances that are commonly known as bacteriocins. The present study aimed to characterize a proteinaceous compound produced by *Leuconostoc lactis* RK18, isolated from a fermented Cambodian food product. It was found that *Ln. lactis* RK18 produced a Class IIa bacteriocin, capable of inhibiting important foodborne pathogens such as *Listeria monocytogenes*. Bacteriocin RK18 remained stable in the temperature range between 30°C to 80°C) and at pH values ranging from 4 to 6. Although susceptible to some detergents, it showed great resistance to several enzymes. RK18 was produced in the early stage of *Ln. lactis* RK18 growth cycle, which brings an advantage when competitive growth is desired or for pathogen inhibition purposes, since it can inhibit pathogenic bacteria before its exponential growth. Even though more studies must be performed, *Ln. lactis* RK18 appears to have the potential to play an important role in the safety of food products and could be a good candidate to be used as a starter culture in the production of other fermented products.

Chapter 4 - Fermented meats are traditional and unique foods which have long been known for centuries. A diverse kind of fermented meat products with different characteristics are produced using widely different formulations, and manufacturing processes involving the nature and tradition of the countries. The microorganisms play a significant role on fermentation process to provide the desirable quality attributes of products. Two major groups of microorganisms participated in meat fermentation include lactic acid bacteria (LAB) and coagulase negative staphylococci (CNS). Moreover, yeasts and molds also involve in the fermentation. Fermented meats are rich sources of protein and fat. They are also the important sources of several other micronutrients such as vitamins (niacin, pyridoxin, thiamine, and riboflavin) and trace elements (zinc, iron, and selenium). However, fermented meats consumption may provide a negative effect on health as a consequence of high levels of fat, cholesterol and low polyunsaturated/saturated fatty acid ratio (PUFA/SFA). Substitution of meat fat with plant-based fat is used to reduce the risk of health disease. Several techniques including addition of dietary fiber, probiotics, synbiotics, acid whey and functional starter cultures are applied to enhance the nutritional values of these products. Fermented meats exhibit several bioactive activities such as antioxidant activity, angiotensin converting enzyme (ACE) inhibitory activity, antidiabetic activity and low cytotoxicity. Therefore, fermented meats are considered as a kind of functional food that can provide a positive effect on human health, particularly when probiotics are used in the manufacturing process. Consumers are aware of safety concerns of fermented meats due to the risk of microbiological, namely foodborne pathogens and chemical hazards especially, biogenic amines, nitrosamines, mycotoxins, and polycyclic aromatic hydrocarbons (PAHs). Starter cultures are employed to ensure the safety of fermented meat products by inhibition of the spoilage and/or pathogenic microorganisms and reduction of the chemical additives particularly, nitrates and nitrites.

Chapter 5 - Fermented shrimp products are traditional fermented foods produced in Southeast Asia (Thailand, Philippines, Vietnam, Indonesia, Myanmar, and Malaysia). *Ka-pi* is a traditional Thai fermented shrimp paste with pinkish or purplish gray to a dark grayish brown-colored, strong odor and salty. On the other hands, *kung-chom* and *kung-som* are other kinds of Thai fermented shrimp paste with orange or red pinkish, sour and salty. Halophilic and lactic acid bacteria predominantly found in fermented shrimp products and play an important role in fermentation. Halophilic and related bacteria consist of various genera: *Alkalibacterium*, *Alkalibacillus*, *Aeromonas*, *Alcaligenes*, *Amphibacillus*, *Bacillus*, *Haloarcula*, *Halorubium*, *Jeotgalicoccus*, *Lentibacillus*, *Macrococcus*, *Micrococcus*, *Oceanobacillus*, *Pseudomonas*, *Salinicoccus*, *Salinivibrio*, *Staphylococcus*, *Sulfidobacillus*, *Vibrio* and *Virgibacillus*, while lactic acid bacteria composed of diverse genera: *Enterococcus*, *Lactobacillus*, *Lactococcus*, *Leuconostoc*, *Pediococcus*, *Sporolactobacillus*, *Tetragenococcus*, and *Weisella*. During fermentation, the protein and fats hydrolysis of shrimp is mediated by the action of microbial enzymes and endogenous enzymes derived from shrimp yield short-chain peptides, free amino acids and short-chain fatty acids which enhance the unique flavor and umami taste and contain biological activities. Additionally, the fermented microorganisms synthesize several metabolites (e.g., alcohol, aldehydes, esters, ketones, organic acids, and others) that contribute to the flavor, aroma, and other beneficially biological activities of fermented shrimp products. Hence, fermented shrimps can be highly nutritious food because of its composition and high digestibility that contribute to health. This chapter display and discuss the production, composition, microorganisms, health benefits, and health risks of fermented shrimp products.

In: Fermented Foods
Editor: Oliver Kovalyov
ISBN: 978-1-53618-656-7

Chapter 1

HYPOVITAMINOSIS B - EXPLOITING LACTIC ACID BACTERIA AS VITAMIN FACTORIES TOWARDS A PROPHYLACTIC APPROACH

Cláudia Maciel, Mónica Ribeiro, Pedro Cruz and Paula Teixeira*
Universidade Católica Portuguesa,
CBQF - Centro de Biotecnologia
e Química Fina – Laboratório Associado,
Escola Superior de Biotecnologia, Porto, Portugal

ABSTRACT

The B group or B-complex vitamins are essential micronutrients possessing physiological effects on a multitude of biological responses, including the immune system. The B-complex vitamins comprise thiamine (B1), riboflavin (B2), niacin (B3), pantothenic acid (B5),

---

* Corresponding Author's Email: pcteixeira@porto.ucp.pt.

pyridoxin (B6), biotin (B7), folate (B9), and cobalamin (B12). Most bacteria, along with plants and fungi, have the ability to biosynthesize *de novo* the vitamins B, whilst humans lack this capacity, hence exogenously acquiring these hydrosoluble molecules from the dietary or microbial sources, namely the endogenous microbiota. Disturbances in the physiological homeostasis of vitamin B elicit a detrimental impact on human health. In this sense, the global upsurge of hypovitaminosis B has been the cause of serious concern, paving the way for the exploitation of suitable alternatives to the conventional chemical-based supplementation. A shifting pattern towards the prevalence of the B-group vitamins deficiencies, previously solely associated with undernutrition and socioeconomic scarcities, has nowadays emerged associated to the renewed vegetarianism-triggered unbalanced diets. In recent years, the development of novel food, pharmaceutical and chemical products have quickly evolved to attend the rising demand for B-group vitamins. Indeed, investigators have concentrated their focus on the exploitation of natural approaches to improve the vitamin B content in food products. The most promising approach is, undeniably, the notable nutraceutical potential of lactic acid bacteria (LAB) as sustainable vitamins biosynthesizers. In the last decades, the development of novel fortified fermented foods, as suitable delivery vehicles to ensure the *in situ* vitamin B bioavailability has been highlighted. These vitamin B bio-enriched products may constitute a feasible prophylactic, or even therapeutic, strategy to counteract hypovitaminosis B. Albeit the significant number of vitamin B-producing LAB identified to date, scarce studies have addressed *in vivo* the impact of the consumption of biofortified functional food products towards the vitamin B deficiency reversion or prevention.

Nonetheless, the scarce bio-based production of LAB relying on intricate metabolic pathways under strict enzymatic regulation poses a prominent obstacle towards the manufacturing scaling-up. Current investigation focuses on genetic bioengineering and integrative multi-omics approaches to generate renewed high-yielding vitamin-producing LAB, and to unveil pivotal enzymes of the metabolic networks underlying the biosynthesis by newly discovered LAB strains.

In this chapter we present a comprehensive overview of the recently discovered, or bioengineered, vitamin B2, B9 and B12-producing LAB and provide in depth analysis of the latest biotechnological applications and potentialities, particularly, the development of novel bio-enriched fermented foods.

1. VITAMIN B2

1.1. Vitamin B2 Discovery – The History of the "Yellow Pigment"

Vitamin B2 was originally described in 1879 by the English chemist Alexander Wynter Blyth, who termed it "lactochrom," referring to a bright yellow compound isolated from cow milk whey (Blyth. A. W. 1879). Nonetheless, it was not until 1934, that Paul Karrer and Richard Kuhn deciphered its chemical structure and proved it by chemical synthesis, describing the isolation of pure, crystalized, lactoflavine (Kuhn, Reinemund, and Weygand 1934; Karrer, Schöpp, and Benz 1935). In 1939, William Sebrell and Roy Butler, evaluating the clinical lesions elicited by a riboflavin deficient diet on a group of patients, demonstrated that human pellagra symptoms were due to the vitamin scarcity, supporting its essential nature as an exogenous food component, concerning human health (Sebrell and Butler 1939). The denomination riboflavin (etymology: from Latin *flavus* for yellow, and "ribo" referring to the ribityl side chain), pointing out that the naturally occurring flavin comprises a univalent radical originated from D-ribitol (the sugar alcohol) in its structure, was adopted in 1952 by the International Commission for the Reform of Biochemical Nomenclature (Wagner-Juaregg 1954).

1.2. Riboflavin Chemistry - Structure and Properties

Riboflavin (7,8-dimethyl-10-(1′-D-ribityl)isoalloxazine) is a water-soluble heterocyclic compound, whose chemical structure, outlined in Figure 1, comprises a tricyclic, nitrogen-containing isoalloxazine ring with a ribityl side chain covalently attached to the nitrogen atom in the position 10 (N-10) of the pyrazine moiety. Indeed, the vitamin is designated as a derivative of dimethylisoalloxazine (7,8-dimethylbenzo[g] pteridine-2,4(3H,10H)-dione), comprising a ribityl substitution, a derivative of the D-ribose moiety of the purine precursor (Abbas and Sibirny 2011).

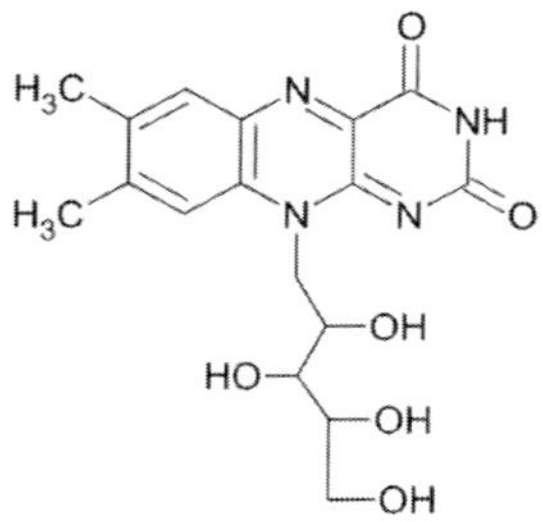

Figure 1. Riboflavin chemical structure (designed using Advanced Chemistry Development, Inc. (ACD)/ChemSketch; ACD/Labs 2019.2.1, Canada).

Pertaining chemical properties and stability, riboflavin is a yellow compound which, following exposure to ultraviolet (UV) light, presents a distinctive yellow-green fluorescence. Aqueous solutions of riboflavin display maximum light absorbances in the UV range, at the wavelengths of 223, 266 and 373 nm, and in visible light at 445 nm (Eitenmiller, Landen Jr., and Ye 2007). Neutral aqueous solutions of the vitamin B2 present the maximum fluorescence emission at 535 nm (Abbas and Sibirny 2011).

Riboflavin isolalloxazine moiety may exist in three distinct oxidation states, namely, the oxidized form (quinone), the semi-quinone form (radical), and fully reduced riboflavin, or hydroquinone state, along with different protonation states, in which the formation of each species relies on the reduction potential employed and pH of the solution, respectively. The vitamin presents an amphoteric character and may occur as neutral, anionic and cationic species in the mentioned redox states (Hemmerich, Veeger, and Wood 1965; Massey 2000; Bailey and Schultz 2016).

Indeed, vitamin B2 is one of the most stable vitamins, having a notable stability in low pH values and being only slightly degraded at neutral pH. The vitamin is scantily soluble in water, presenting the lowest hydrosolubility amongst the B group vitamins (Gropper, Smith, and Groff 2009). In order to circumvent this scarce water solubility, the sodium salt of riboflavin 5'-phosphate is commonly used to prepare aqueous solutions (Berry Ottaway 2010). Nonetheless, riboflavin is soluble in alkaline solution, albeit turning out unstable under exposure to high pH conditions (Beinert 1960; Abbas and Sibirny 2011).

Riboflavin is a heat-stable vitamin, capable of withstanding high temperatures, independently of the heat source. Indeed, a significantly high decomposition temperature was documented, namely, 271-293°C (Wagner-Juaregg 1954), and the vitamin is stable under the different processing methods in which the temperature is applied, namely microwaves, infra-red radiation, high-pressure steam, or hot air convection (Cardoso, Libardi, and Skibsted 2012).

Albeit the riboflavin thermal stability, photolysis elicits its prompt decomposition, originating derivatives possessing ribityl side chains completely or partly degraded; lumiflavin (7,8,10-trimethylisoalloxazine) is synthetized in alkaline solutions, whilst lumichrom (7,8-dimethylisoalloxazine) is accumulated at acidic and neutral pH values. The light exposure also originates two riboflavin excited states (singlet and triplet; ^{1}Rib* and ^{3}Rib*, respectively), which will in turn react with other compounds present in the matrices, resulting in their degradation (Cardoso, Libardi, and Skibsted 2012).

Riboflavin originates water-insoluble hydrophobic complexes (precipitates) with salts of metal ions, e.g., magnesium, copper, iron, cobalt, nickel and mercury (Berezovskii 1973; Abbas and Sibirny 2011).

The oxygen appears not to impact riboflavin stability during processing, although, throughout storage, the interaction of the previously mentioned excited states with oxygen leads to riboflavin oxidation (Burch 2011; Schiano et al. 2019).

1.3. Biochemical Pathway of Riboflavin Biosynthesis

1.3.1. Riboflavin De Novo Biosynthetic Pathway

Most bacteria, along with all plants and fungi present the capability to biosynthesize *de novo* riboflavin, being a notable source of this vitamin for animals, particularly humans.

The riboflavin biosynthesis route has been investigated in Gram-positive, along with Gram-negative bacteria, albeit more thoroughly in *Bacillus subtilis* (Perkins and Pero 2002) and *Escherichia coli* (Bacher,

Eberhardt, and Richter 1996; Bacher et al. 2000). The pathway encompasses a seven-step enzymatic network, in which a four-gene operon *rib*GBAH encodes the enzymes responsible for the catalytic conversion of the metabolic precursors, one molecule of 5'-guanosine triphosphate (GTP) and two molecules of ribulose-5'-phosphate (R5P), proceeding through pyrimidines and pteridine intermediates towards riboflavin biosynthesis (Perkins and Pero 2002). Lactic acid bacteria (LAB), namely, *Lactococcus lactis* subsp. *cremoris* NZ9000, *Leuconostoc mesenteroides* ATCC 8293 and *Pediococcus pentosaceus* ATCC 25745, harbor the abovementioned single riboflavin operon (*rib* operon), comprising the complete set of riboflavin biosynthetic genes in the same order as *B. subtilis* (Vitreschak et al. 2002b; Capozzi et al. 2012).

Noteworthy, the genes order comprised in the *rib*GBAH operon is not correlated with the pathway enzymatic reactions order. Nevertheless, and in accordance with comparative genome evaluation, it was documented that the genetic information for the vitamin B2 synthesis may be associated with species or strain-specific features. In fact, the whole genome sequencing of *Lactobacillus plantarum* strain WCFS1 unveiled an incomplete operon, devoid of the entire *rib*G gene and comprising a truncated copy of the *rib*B gene (Kleerebezem et al. 2003).

The gene *rib*G encodes the bifunctional pyrimidine deaminase-reductase enzyme, which comprises two domains, namely, the C-terminal located deamination domain (diaminohydroxyphosphoribosyl-aminopyrimidine deaminase (EC 3.5.4.26)) and the reductase domain (EC 1.1.1.193), located in the N-terminal region. The fused gene *rib*A encodes a bifunctional enzyme, namely, GTP cyclohydrolase II and 3,4-dihydroxy 2-butanone 4-phosphate synthase. The gene *rib*H encodes riboflavin synthase beta-subunit, comprised in an enzyme complex, consisting of β-subunits (lumazine synthase). The mentioned complex also contains α-subunits specified by the gene *rib*B, which encodes the riboflavin synthase alpha-subunit (Bacher et al. 2000). Additionally, to the polycistronic *rib* operon, the presence of the monocystronic gene *rib*C encoding the bifunctional flavokinase/FAD synthetase indicates that the riboflavin may be biochemically converted to the biologically active forms of its derivatives coenzymes, flavin adenine dinucleotide (FAD) and flavin mononucleotide (FMN), in order to ensure its metabolic function (Figure 2). Another *trans*-acting gene, *rib*R encoding the monofunctional riboflavin kinase, along with the gene *rib*C, is not genetically linked to the biosynthetic riboflavin operon (Abbas and Sibirny 2011).

The imidazole ring of guanosine 5' triphosphate is hydrolytically opened through the catalysis of the enzyme GTP hydrolyse II (EC 3.5.4.25, RibA), eliciting the release of its C(8) (formamide) as formate (formyl group), concomitantly with inorganic pyrophosphate. The enzymatic product 2,5-diamino-6-ribosylamino-4(3*H*)-pyrimidinone- 5' phosphate is subsequently deaminated through the hydrolytic cleavage of the position 2 amino group of the heterocyclic ring, which is catalysed by the bifunctional enzyme riboflavin-specific deaminase/reductase (RibG), generating 5-amino-6-(ribosylamino)- 2,4 (1*H*, 3*H*) – pyrimidinedione-5' phosphate. The following reduction of the phosphoribosyl side chain of the latter compound through the catalysis of riboflavin-specific deaminase/reductase (RibG) originates 5-amino-6-(ribitylamino)-2,4 (1*H*, 3*H*) – pyrimidinedione-5' phosphate. Subsequently, a dephosphorylation of the latter compound occurs (which mechanism remains elusive) generating 5-amino-6-ribitylamino-2,4(1*H*, 3*H*)-pyrimidinedione, which, condensed with the 4-carbon compound, 3,4-dihydroxy-2-butanone 4-phosphate by lumazine synthase (beta-subunit) (RibH), originates the bicyclic pteridine 6,7-dimethyl-8-ribityllumazine. The carbohydrate type substrate of the RibH enzyme is derived from ribulose-5' phosphate in a reaction catalyzed by the enzyme GTP hydrolyse II/3,4-dihydroxy-2-butanone-4- phosphate synthase (RibA), originating 3,4-hydroxy-2-butanone 4-phosphate. The final step of the biosynthetic pathway comprises the dismutation of 6,7-dimethyl-8-ribityllumazine catalyzed by riboflavin synthase (alpha-subunit, RibB). Stoichiometrically, the riboflavin synthesis requires one equivalent of GTP and two equivalents of ribulose 5-phosphate (Bacher et al. 2000; Burgess et al. 2004; Abbas and Sibirny 2011; Haase et al. 2014). The riboflavin biosynthetic pathway is depicted in Figure 3.

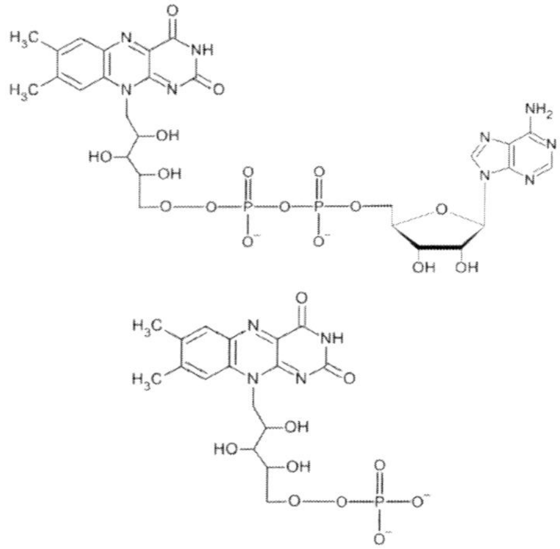

FAD

FMN

Figure 2. Flavin adenine dinucleotide (FAD) and flavin mononucleotide (FMN) chemical structure (designed using ACD/ChemSketch; ACD/Labs 2019.2.1).

1.3.2. Regulation of Riboflavin Biosynthesis

The regulation of riboflavin biosynthesis predominantly occurs at the stage of its biosynthetic enzymes' synthesis and activity, being the principal effector of the mechanism the flavin coenzyme, FMN (Abbas and Sibirny 2011). In this sense, the regulatory mechanism underlying riboflavin biosynthesis pathway encompasses a metabolite-sensing riboswitch, performing at transcriptional and translational level (Vitreschak et al. 2002b; García-Angulo 2017). The riboswitch responsible for the vitamin B2 synthesis modulation is termed the FMN-recognizing riboswitch, a messenger RNA (mRNA) element (structure/sequence) commonly comprised within the 5' untranslated region of mRNA, localized upstream the polycistronic riboflavin (*rib*) operon and monocystronic *rib* genes (Lee, Blount, and Breaker 2009; García-Angulo 2017). These regulatory elements encompass two distinct structural domains, namely, one sensor domain termed aptamer (RFN element), a receptor to which the small molecule FMN tightly binds with high selectivity, and an expression platform, endowed with the capability to be allosterically altered, eliciting the modulation of the downstream coding sequences (Vitreschak et al. 2002a; Nudler and Mironov 2004; Mehdizadeh Aghdam, Hejazi, and Barzegar 2016; Pedrolli et al. 2015).

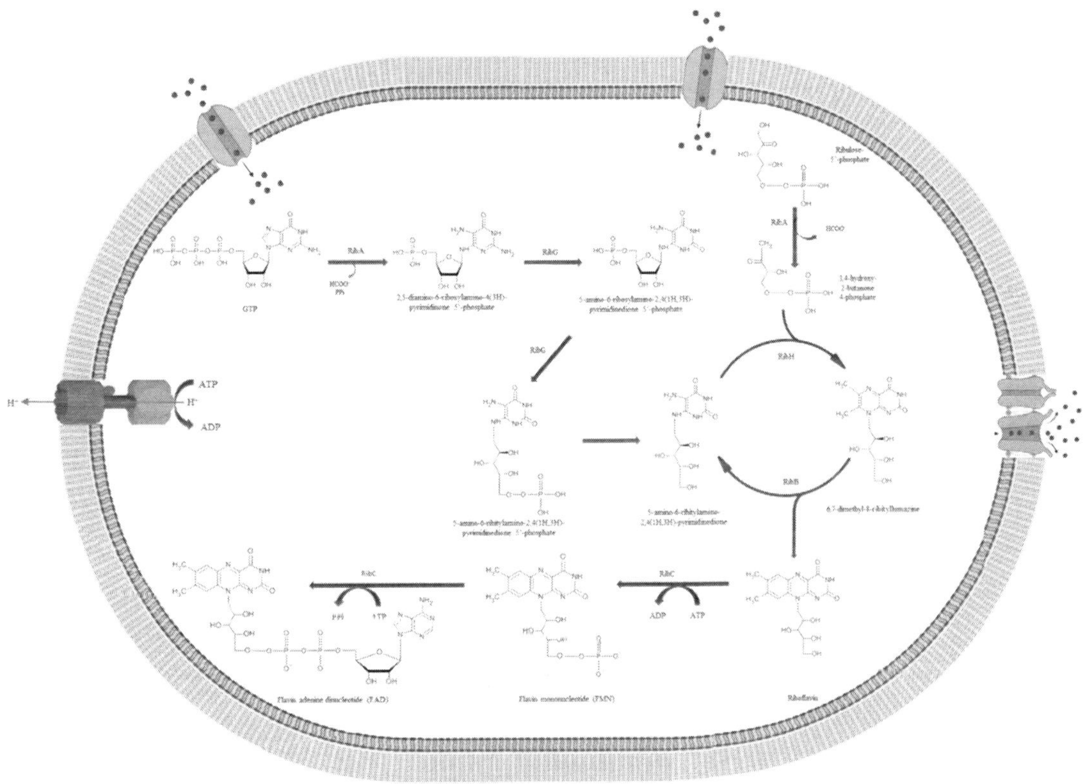

Figure 3. Riboflavin and flavocoenzymes *de novo* biosynthetic pathway (designed using ACD/ChemSketch; ACD/Labs 2019.2.1, adapted from Abbas and Sibirny (2011)).

Flavin mononucleotide entrapment in the aptamer region triggers a secondary structure alteration of the expression domain of the FMN riboswitch. The switching sequence constitutes a common region to both aptamer domain and expression platform and, once embedded in one of those domains, determines in which of the two secondary structures (terminator/antiterminator stem-loop) the expression domain folds into, and hence dictates the expression fate of the mRNA (Edwards and Batey 2010; Tucker and Breaker 2005).

The FMN riboswitch may modulate transcription or translation of mRNA of riboflavin operon, encoding the enzymes involved in the biosynthesis of riboflavin, through the generation of a terminator hairpin or sequestering the ribosome binding site (via base pairing) (Krajewski, Ignatov, and Johansson 2017).

Once the intracellular FMN concentration surpasses the threshold level, binding to the riboswitch aptamer originates a structural rearrangement, in which the switching sequence is incorporated into the aptamer domain, eliciting a conformational alteration in the secondary structure of the adjacent region, the regulatory platform (Garst and Batey 2009; Mehdizadeh Aghdam, Hejazi, and Barzegar 2016). This modification allosterically remodels the expression platform, which results in the formation of a hairpin, which may be a terminator or sequester hairpin. In this sense, transcriptional repression occurs through a termination-antitermination mechanism, owing to the formation of the aforementioned terminator stem-loop, which signals RNA polymerase to prematurely cease transcription. In opposition, once FMN is lacking or the intracellular level is scarce, the expression platform integrates the switching sequence folding into an antiterminator hairpin (secondary structure), which permits the normal transcription of the *rib* operon (Vitreschak et al. 2002b).

Translational modulation (repression) is also mediated by a similar mechanism, when the hairpin generated from the alteration of secondary structure of RFN is a sequester hairpin, which sequesters/masks the Shine-Dalgarno (SD) sequence (ribosome binding site) through the formation of a loop, resulting in impairment of the binding of the ribosome to the mRNA and consequent translational repression (Nudler and Mironov 2004; Mehdizadeh Aghdam, Hejazi, and Barzegar 2016; Serganov and Nudler 2013).

Whilst Gram-negative bacteria possess one FMN riboswitch (SA1; *rib*GBAH), responsible for regulation of riboflavin *de novo* biosynthesis, Gram-positive bacteria possess an additional FMN riboswitch (SA2; *rib*U), which controls the riboflavin uptake and, therefore, are also capable of using external riboflavin (if available) (Krajewski, Ignatov, and Johansson 2017).

One other putative regulation mechanism of the vitamin B2 synthesis is associated with the modulation of the biosynthetic enzyme GTP cyclohydrolase II activity through feedback inhibition (Abbas et al. 2011). Albeit Hümbelin and co-workers (1999) documented that GTP cyclohydrolase II is a rate-limiting enzyme in the riboflavin synthesis of *B. subtilis*, there is scarce knowledge concerning the allosteric feedback inhibition of this enzyme.

1.4. Industrial Production of Riboflavin

The prominent interest in vitamin B2 was the driving force which prompted riboflavin chemical synthesis (Karrer, Schöpp, and Benz 1935; Kuhn 1936). The main chemical route encompasses a multi-step procedure initiated either from D-glucose or D-ribose, which is converted into arabonate through oxidation. Thereafter, the latter compound is epimerized to ribonate and converted into ribonolactone, which is electrochemically reduced to D-ribose, employing amalgam. The exogenous supply of xylidine elicits the formation of ribityxylidine, which, further associated with an aniline derivative, generates phenylazo-ribitylxylidine. The final reaction of the chemical synthesis comprises the cyclocondensation of phenylazo-ribitylxylidine with barbituric acid, originating riboflavin as the final product (Wolf et al. 1983; Schwechheimer et al. 2016).

Nonetheless, currently, the riboflavin industrial production has transitioned from the chemical synthesis to be solely performed through the fermentative process, owing to the higher economic and ecological suitability. This shifting pattern also occurred due to the remarkable progresses of genetic and bioprocesses engineering. Indeed, the two major fermentation processes are accomplished through the employment of the bioengineered derivatives of the Gram-positive bacterium *B. subtilis* and filamentous hemiacomycete *Ashbya gossypii*. High-performing strains may be generated by a conjunction of classical mutagenesis, genome shuffling and metabolic engineering (Schwechheimer et al. 2016). The industrially produced riboflavin is intended for dietary supplementation, animal feed, and as an additive for the food industry.

A multitude of other interesting microorganisms possess the capability to overproduce riboflavin due to intrinsic mutations within FMN riboswitch, namely, the lactic acid bacteria (*Lc. lactis*, *L. plantarum* and *Lc. mesenteroides*), as well as *Propionibacterium freudenreichii* (Burgess et al. 2006; Sybesma et al. 2004), which will be more thoroughly scrutinized further ahead.

1.5. Metabolic Engineering – Towards Riboflavin Overproduction

Two distinct approaches have been exploited to optimize bacterial riboflavin biosynthesis, namely, selection of roseoflavin-resistant spontaneous mutants and target metabolic bioengineering.

Contrarily to *A. gossypii*, which naturally oversynthesizes its pseudo-secondary metabolite (riboflavin) during sporulation as a mechanism of protection and detoxification, *B. subtilis* is not regarded as a spontaneous riboflavin overproducer (Stahmann et al. 2001; Jiménez et al. 2005; Aguiar, Silva, and Domingues 2015). Nonetheless, the food industry exploitation of an approach relying on a chemical analog of the vitamin B2 (roseoflavin) to screen for, and select, riboflavin-overproducing bacterial strains appears to be promising (Jayashree et al. 2011; Thakur, Tomar, and De 2016).

As previously detailed, the regulation of riboflavin biosynthesis occurs through turning off transcription of the riboflavin biosynthetic genes, or translation, following the FMN binding to FMN riboswitches, which impairs the vitamin synthesis. Indeed, when mutations occur in the *de novo* synthesis genes operon (*rib*GBAH) or riboflavin transporter (*rib*U) there is no ligand binding and consequent production shutdown, and riboflavin is continuously synthetized (overproduction) (Lee, Blount, and Breaker 2009).

Roseoflavin (8-demethyl-8-dimethylamino-riboflavin) is a riboflavin and FMN natural structural analogue (Otani, Kasai, and Matsui 1980) which, albeit being biosynthesized by the filamentous Gram-positive bacteria *Streptomyces davawensis* and *Streptomyces cinnabarinus*, nowadays it is solely (exclusively) chemically produced. Roseoflavin uptake occurs *via* riboflavin transporters and once inside the bacterial cell it is converted into roseoflavin mononucleotide (RoFMN) and roseoflavin adenine dinucleotide (RoFAD), the analogs of the cofactors, FMN and FAD, respectively.

Roseoflavin possesses antimicrobial activity as, by binding to both SA1 and SA2 FMN riboswitches, signals the turning off of riboflavin production by both mechanisms (*de novo* and uptake), leading to cellular death. Owing to this feature, roseoflavin is used in the food industry to identify riboflavin-overproducing spontaneous mutants. The rationale is that bacterial cells harboring mutations in *rib*GBAH or *rib*U FMN riboswitches will resist exposure to roseoflavin, since this chemical compound is not able to bind, which will also succeed with riboflavin, leading to an absence of repression of *rib* operon expression and consequent overproduction (Blount and Breaker 2006; Lee, Blount, and Breaker 2009). In fact, Blount and Breaker (2006) and co-workers (2009) documented mutations within the aptamer domain of the *rib* operon FMN riboswitches in roseoflavin-resistant, riboflavin overproducing, strains.

A myriad of spontaneous roseoflavin-resistant LAB and propionibacteria strains, overproducing riboflavin, have been unveiled, which may not be considered genetically modified organisms. Nevertheless, in comparison to the industrial riboflavin microbial producers, namely, *A. gossypii* (possessing a titer of 26 to 30 g L^{-1}) (Liu et al. 2020), the evaluated mutant strains demonstrated a scarce vitamin synthesis, specifically, several thousand-fold lower than the industrial yields.

Sybesma and colleagues (2004) documented that a roseoflavin-resistant mutant *Lc. lactis* CB010 yielded extracellular riboflavin content greater than 1.2 mg L^{-1}. Burgess and co-workers (2006), evaluating the vitamin content of riboflavin-overproducing food grade microorganisms, unraveled that *P. freudenreichii* B2336 was capable to produce over 3 mg L^{-1}, whilst *Lc. mesenteroides* CB207 and *L. plantarum* CB300 synthetized approximately 0.55 and 0.65 mg L^{-1}, respectively.

Concerning metabolic engineering, the different approaches to generate riboflavin-overproducing strains are based on genetic manipulation of the riboflavin biosynthetic pathway, of the purine biosynthetic route (which originates the precursor GTP), of the pentose phosphate pathway, the production of amino acid serine (the precursor of the purines synthetic route), amongst others (Liu et al. 2020). With respect to the first strategy, Burgess and colleagues (2004) performed overexpression of the *rib*GBAH operon in *Lc. lactis* NZ9000 upon nisin induction, achieving a riboflavin concentration of 24 mg L^{-1}. Regarding the overexpression of the purine biosynthetic pathway, Jayashree and co-workers (2011) improved in approximately 50% the vitamin B2 production by *L. fermentum* GKJFE via the inactivation of the *fol*E gene, encoding the GTP cyclohydrolase I, which consequently elicited enhanced number of GTP molecules available for the riboflavin synthesis.

1.6. Riboflavin Human Absorption

Humans lack the capability of synthesizing riboflavin, hence acquiring the vitamin from the dietary sources, through intestinal absorption (Foraker, Khantwal, and Swaan 2003). In fact, the intestine presents a pivotal role concerning the regulation of the physiological homeostasis of the vitamin B2, being exposed to two distinctive riboflavin sources, namely, the dietary and the endogenous microbiota (Lakhan, Subramanian, and Said 2017).

Nonetheless, the dietary riboflavin is predominantly found in the form of the flavin coenzyme FAD and, to a lesser extent, the mono-phosphorylated form, FMN, associated with food proteins (enzymes), requiring hydrolysis to generate free riboflavin (Balasubramaniam, Christodoulou, and Rahman 2019; Institute of Medicine and Food and Nutrition Board 1998).

Following ingestion, and once in the stomach, the exposure to the harsh acidic conditions provided by the gastric acids, elicits the liberation of the non-covalently bound coenzymes FAD and FMN from the enzymes. Thereafter, FAD and FMN are promptly hydrolysed to free riboflavin in the upper small intestine by non-specific phosphatases of the brush border and basolateral membranes of the enterocytes, which is paramount for the vitamin absorption process (Akiyama, Selhub, and Rosenberg 1982; Daniel, Binninger, and Rehner 1983; Powers 2003).

The absorption of free riboflavin, which is the only absorbable form, primarily occurs in the proximal small intestine via a saturable, carrier-mediated, transport process (by RFVT3) with high specificity (Balasubramaniam, Christodoulou, and Rahman 2019; Foraker, Khantwal, and Swaan 2003). Nonetheless, vitamin B2 is also absorbed in other organs, namely, duodenum, colon and rectum, and also through the same type of transport system, mediated by the same transporter (RFVT3) (EFSA 2017; Mosegaard et al. 2020). It has been documented that, besides RFVT3, which is the predominant riboflavin transporter, RFVT1 and RFVT2 are also present in the gastrointestinal tract. The RFVT1, RFVT2 and RFVT3 transporters are encoded by the *SLC52A1*, *SLC52A2* and *SLC52A3*, respectively (Balasubramaniam et al. 2019; Mosegaard et al. 2020; Subramanian et al. 2015, Lakhan et al. 2017).

Afterwards, riboflavin may be converted into FMN and FAD by catalytic enzymes in the enterocytes, or transported by RFVT1 and RFVT2, located in the basolateral membrane of the epithelial cells of the gastrointestinal tract, to the bloodstream for distribution to the tissues

(EFSA 2017; Mosegaard et al. 2020; Balasubramaniam, Christodoulou, and Rahman 2019).

The riboflavin transported in the blood plasma is weakly associated to albumin or immunoglobulins, or is converted into FMN and FAD in leucocytes or erythrocytes (Innis et al. 1986; Balasubramaniam et al. 2019; Institute of Medicine and Food and Nutrition Board 1998).

The vitamin uptake to the organs' cells, at physiological concentrations, may need the intervention of a specific carrier to facilitate the transport (Institute of Medicine and Food and Nutrition Board 1998). The highest concentrations of the vitamin B2 are found in the liver, heart and kidney (Monteiro and Perrone 2012).

The vitamin which is not absorbed is excreted in the urine in one of two forms, namely, as free riboflavin, or a derived metabolite (7-hydroxymethylriboflavin and lumiflavin), and it has been suggested that, from the riboflavin absorbed, only a part is present in the plasma, since the liver sequesters the remaining portion upon the first passage in the portal vein from the intestine (Chastain et al. 1987; Mosegaard et al. 2020).

Apart from being acquired by dietary intake, riboflavin may also be biosynthesized in the gut by endogenous microbiota (Hill 1997). Magnúsdóttir and colleagues (2015) explored the presence of biosynthetic pathways of B-vitamins (among which was B2) in 256 common human gut bacteria by studying the corresponding genome annotations. The study revealed that 65% of the evaluated bacteria presented the genetic composition that enables riboflavin production, which is in accordance with the previously reported by Hill (1997) which, as abovementioned, found gut bacteria to be a source of group B vitamins.

1.7. Biological Activity in Humans

Riboflavin is crucial for the maintenance of human health, possessing a pivotal role concerning cellular functions (metabolism, proliferation, growth and survival) (Lakhan, Subramanian, and Said 2017).

In order to be biochemically active and accomplish its pivotal metabolic function in the living cell, since no human enzyme presents the capability to utilize riboflavin *per se*, the vitamin must be chemically converted into the biologically active forms of its derivatives known as flavin coenzymes, namely riboflavin-5'-phosphate FMN and FAD (Rivlin and Pinto 2001; Lienhart, Gudipati, and Macheroux 2013). Indeed, riboflavin phosphorylation at the position 5' of the ribityl chain, triggered by the catalysis of the enzyme riboflavin kinase (EC 2.7.1.26), originates riboflavin-5'-phosphate, the FMN coenzyme. Flavin adenine dinucleotide is biosynthesized in a succeeding reaction of FMN adenylylation catalysed by FAD synthase or FMN adenylyltransferase (EC 2.7.7.2), which transfers adenylyl moieties from ATP to FMN (Rivlin and Pinto 2001).

Owing to the chemical versatility, those essential cofactors of flavoenzymes are involved in a plethora of metabolic pathways, essential biological oxidation-reduction reactions encompassing lipid, carbohydrates and amino acid metabolism, acting as electron carriers (mediating the electron transfer), and in the synthesis of some vitamins and coenzymes (Massey 2000; Abbas and Sibirny 2011).

The flavin coenzymes, FAD and FMN, are the redox cofactors of a myriad of flavoproteins, namely, dehydrogenases, oxidases and reductases. These flavoenzymes may transfer one or two electrons, in the form of hydrogen atoms (electron and proton) and hydride ions (the equivalent of a proton and two electrons), promoting, for instance, the substrates dehydrogenation in a two-electron reduction reaction (Massey 2000).

The mentioned flavoprotein coenzymes play a pivotal role in oxidative metabolism, participating as a one- or two-electron shuttles to numerous metal centers *via* their free radical states, constituting moieties of the multi-redox-center enzymes, namely, nicotinamide adenine dinucleotide (NADH) dehydrogenase (containing FMN as prosthetic group, complex I of the mitochondrial respiratory pathway), succinate dehydrogenase (comprising FAD as the prosthetic group, complex II in the electron transport chain and citric acid cycle), xanthine oxidase/dehydrogenase (bounded FAD, ferredoxin iron-sulphur clusters, purines catabolism), nitric oxide synthase (encompassing FAD and FMN as coenzymes, along with three other cofactors, amongst which the heme group is the electron acceptor of FMN) (Massey 2000) and microsomal NADPH-cytochrome P450 reductase (CPR), comprising FAD and FMN as prosthetic groups which mediate the electron transfer between pyridine nucleotide and heme redox centre (Massey 2000; Wang et al. 1997).

Most human flavoproteins with prominent physiological functions utilize FAD (84%) as cofactor (Lienhart, Gudipati, and Macheroux 2013), namely the mitochondrial fatty acid β-oxidation, in which the flavoprotein acyl-coA dehydrogenase (ACAD) use FAD as cofactor (Kim, Wang, and Paschke 1993; Swigoňová, Mohsen, and Vockley 2009); amino acids oxidative deamination, mediated by the dimeric flavoprotein L-amino acid oxidase (LAAO), encompassing non-covalently attached FAD (Curti, Ronchi, and Simonetta 1992); and the oxidative decarboxylation of pyruvate to acetyl CoA mediated by pyruvate dehydrogenase complex which encompasses three main catalytic components, of which the dihydrolipoyl dehydrogenase enzyme comprises a FAD binding domain (Mande et al. 1996; Patel et al. 2014).

As previously mentioned, riboflavin is associated to the metabolic pathways of some B-vitamins, namely, in the conversion of vitamins B6, B9 and B12 into their active forms, owing to the fact that the prosthetic groups FAD or FMN are comprised in the corresponding metabolizing enzymes (Mushtaq et al. 2009; Renz 2007; McKinley 2000; Lienhart, Gudipati, and Macheroux 2013). Indeed, the enzyme pyridoxamine phosphate oxidase (PNPO, EC 1.4.3.5) is dependent of FMN, being engaged in the conversion of pyridoxine 5´phosphate and pyridoxamine towards the coenzyme pyridoxal 5'-phosphate and is located in various tissues, namely erythrocytes (Mushtaq et al. 2009). Moreover, concerning cobalamin, FMN is the metabolic precursor of dimethylbenzimidazol moiety of coenzyme B12 (Renz 2007).

It has also been documented that the vitamin B2 is pivotal concerning the function of enzymes associated with the homocysteine metabolism, owing to the fact that the flavoenzyme methylenetetrahydrofolate reductase (MTHFR) requires FAD as prosthetic group (Bates and Fuller 1986), and hence riboflavin status might impact the homocysteine metabolism.

MTHFR (EC 1.8.1.7) is a crucial enzyme in the folate cycle, catalyzing the reduction of N^5, N^{10}-methylenetetrahydrofolate to N^5-methyltetrahydrofolate (Jacques et al. 2002). The latter compound is responsible for the donation of a methyl group in the remethylation of homocysteine into methionine mediated by methionine synthase. Concerning the mentioned homocysteine metabolism, the second pathway (trans-sulfuration) is also dependent on riboflavin. Once trans-sulfuration of homocysteine occurs, homocysteine is irreversibly condensed with serine to originate cystathionine catalyzed by cystathione-synthase (Mudd, Levy, and Skovby 1995; McKinley 2000). Riboflavin deficiency may promote higher concentrations of plasma homocysteine, which has been correlated with a higher risk of cardiovascular disease (Powers 2003; Abbas and Sibirny 2011).

The antioxidant potential of riboflavin, as metabolic precursor of FAD, which is a coenzyme for the glutathione reductase and xanthine oxidase systems, has been demonstrated, owing to the notable capability to diminish the cellular oxidative stress (Lakhan, Subramanian, and Said 2017). Indeed, the antioxidant enzyme, glutathione reductase (Gor), relies on the cofactor FAD to catalyze the reduction of the oxidized form glutathione disulfide (GSSG) to the sulfhydryl form glutathione (GSH) through the utilization of NADPH, a crucial step in protecting cells from oxidative damage, since GSH is a notable scavenger of free radicals and reactive oxygen species (ROS), and preserving the cellular reducing environment (Deponte 2013; Rivlin 2001). Pertaining to individuals which possess a glucose-6-phosphate dehydrogenase (G6PD) deficiency, the most common enzyme disorder originated by an enzyme defect, glutathione reductase presents a higher requirement for the prosthetic FAD, eliciting an elevated *in vitro* enzymatic activity (Flatz 1970).

The prominent anti-inflammatory features of riboflavin enable the vitamin to undertake a crucial role concerning normal immunity function (Liu and Zempleni 2014), along with preserving the regular physiological intestinal homeostasis (Nakano et al. 2011). Moreover, a participation in the protein folding in the endoplasmic reticulum was documented (Tu et al. 2000).

1.8. Dietary Requirements in Humans

Riboflavin deficiency is commonly a consequence of insufficient dietary intake, resulting from limited access to appropriate food products (FAO and WHO 2004), although it may also derive from deficient dietary options or vitamin malabsorption. Owing to the clinical manifestations of such deficiency, which may range from mild (sore throat) to severe (anemia) (Institute of Medicine and Food and Nutrition Board 1998; EFSA 2017), the necessity to define/establish reference values of dietary intake daily requirements emerged. Several distinct institutions worldwide have dedicated efforts to understand and quantify those necessities. Based on scientific evidence (reports, research articles and similar) and examining different aspects, namely physiology, metabolism, distribution to tissues, excretion, bioavailability in food products, populations with specific needs, amongst others, riboflavin requirements were determined. Nonetheless, since distinct assumptions were considered by different institutions, although similar to some extent, the established requirements diverge between institutions.

The Standing Committee on the Scientific Evaluation of Dietary Reference Intakes, Food and Nutrition Board, of The Institute of Medicine (Institute of Medicine and Food and Nutrition Board 1998) defined several dietary reference intakes (DRIs), among which were the estimated average requirement (EAR) and the recommended dietary allowance (RDA). The EAR is the estimated daily intake, intended to cover the need of 50% of the population, and RDA, which is based in EAR, is the average daily intake intended for 97-98% of the population to meet the required amount of the nutrient. One other DRI, namely, adequate intake (AI), was also established, when EAR was not known, and was expected to exceed EAR. In the report, concerning children under 12 months, it was proposed to use AI instead of EAR.

At the time that the report was elaborated, the committee assumed that there were differences between life stages (infants, children and adolescents, adults and elder) and genders (only above 9 years-old), concerning riboflavin requirements, and also considered that pregnancy and lactation were life stages during which requirements were specific.

The elderly were considered a particular group since there was an apparent correlation between lower fat to carbohydrate ratio and a decrease in riboflavin requirement regarding this specific population. In this sense, the requirement for that age group was lower.

Table 1. Riboflavin dietary reference intakes (estimated average requirement, EAR; recommended dietary allowance, RDA; adequate intake, AI), according to age, gender and population group (Institute of Medicine and Food and Nutrition Board 1998)

Age	Gender	EAR (mg day^{-1})	RDA (mg day^{-1})
0-6 months	-	0.3 (~ 0.04 mg kg^{-1}) (AI)	-
7-12 months	-	0.4 (~ 0.04 mg kg^{-1}) (AI)	-
1-3	-	0.4	0.5
4-8	-	0.5	0.6
9-13	Boys	0.8	0.9
14-18		1.1	1.3
9-13	Girls	0.8	0.9
14-18		0.9	1.0
19-30	Men	1.1	1.3
31-50		1.1	1.3
51-70		1.1	1.3
19-30	Women	0.9	1.1
31-50		0.9	1.1
51-70		0.9	1.1
>70	Men	1.1	1.3
>70	Women	0.9	1.1
14-18	Pregnancy	1.2	1.4
19-30		1.2	1.4
31-50		1.2	1.4
14-18	Lactation	1.3	1.6
19-30		1.3	1.6
31-50		1.3	1.6

Moreover, pregnancy and lactation were also regarded as distinct groups due to the intrinsic relationship between the mothers and the fetus, newborn, and breastfeeding infants, and therefore, requirements were increased by 0.3 mg day^{-1} (pregnancy) and 0.4 mg day^{-1} (lactation).

Consequently, different riboflavin dietary reference intakes (EAR, RDA and AI) were defined according to the age, gender and population group, which are detailed in Table 1.

A joint Food and Agriculture Organization/World Health Organization (FAO/WHO) Expert Consultation on Human Vitamin and Mineral requirements also elaborated a report (FAO and WHO 2004), in which requirement values for dietary intake of riboflavin were established and, in that case, recommended nutrient intakes (RNIs) were the equivalent to RDA. Nonetheless, some different considerations were contemplated when establishing the requirements, in comparison with the American report of the Institute of Medicine (Institute of Medicine and Food and Nutrition Board 1998). In the FAO/WHO report, the riboflavin requirements regarding adolescents were not segmented (10-18 years), and the elderly were not considered as having specific recommended intakes. The RNI values for each group are listed in Table 2, and it was observed that comparatively to those reported in the American document of the Institute of Medicine, were slightly higher.

Table 2. Riboflavin recommended nutrient intakes (RNIs), by group (FAO and WHO 2004)

Group	RNIs (mg day^{-1})
Infants and children	
0-6 months	0.3
7-12 months	0.4
1-3 years	0.5
4-6 years	0.6
7-9 years	0.9
Adolescents	
Females 10-18 years	1.0
Males 10-18 years	1.3
Adults	
Females, 19+ years	1.1
Males, 19+ years	1.3
Pregnant women	1.4
Lactant women	1.6

More recently, in 2017, an European report was released, in which European Food Safety Authority (EFSA) Panel on Dietetic Products, Nutrition and Allergies (NDA) (EFSA 2017) established riboflavin dietary reference values. The Panel presented different assumptions regarding some aspects, namely the inexistence of differences between genders and also between adults of different ages (elderly included), and different age-intervals were assumed, in which adolescence was segmented into two distinct periods. No requirements were determined for infants younger than seven months since the report assumed that riboflavin intake through breastmilk ingestion is sufficient. The requirements were determined in terms of average requirements (ARs), similar to RDA and RNIs, and population reference intakes (PRIs), which are derived from ARs, assuming a 10% coefficient of variation. Riboflavin requirements were slightly higher than the mentioned in the other reports, and are depicted in Table 3.

Table 3. Riboflavin dietary reference values (average requirement, AR; population reference intake, PRI) (EFSA 2017)

Age	AR (mg day^{-1})	PRI (mg day^{-1})
7-11 months	-	0.4[(a)]
1-3 years	0.5	0.6
4-6 years	0.6	0.7
7-10 years	0.8	1.0
11-14 years	1.1	1.4
15-17 years	1.4	1.6
≥18 years	1.3	1.6
Pregnancy	1.5	1.9
Lactation	1.7	2.0

[(a)] Adequate Intake.

1.9. Riboflavin Deficiency (Etiology, Pathophysiology and Epidemiology)

Owing to the previously detailed paramount vitamin B2 role in human health, it is expectable that disturbances in the physiological riboflavin homeostasis elicit a detrimental health impact. Indeed, riboflavin is an essential vitamin to humans and its deficiency is designated ariboflavinosis.

Ariboflavinosis may be originated by a scarce dietary intake or endocrine anomalies. Likewise, hemodialysis, hepatic disorder, chronic alcoholism (Neville et al. 1968; Rosenthal et al. 1973) and diabetes mellitus (Kodentsova et al. 1994) may also lead to riboflavin deficiency.

The riboflavin deficiency has long been investigated given its association with several pathologies. The vitamin B2 hypovitaminosis may be manifested in a multitude of clinical symptoms, some of which may be mild, for instance, hair loss, dermatitis and sore throat, whilst others may be more severe, e.g., glossitis and impaired nerve functions. Other pathologies also associated with low levels of vitamin B2 include changes in mucous membranes, preeclampsia, visual impairment, anemia and elevated homocysteine concentrations (DiBaise and Tarleton 2019; Mosegaard et al. 2020; Titcomb and Tanumihardjo 2019; Shi et al. 2014). A few studies have also demonstrated an association between low riboflavin levels and prognosis, development, survival and relapse of esophageal squamous cell carcinoma (Li et al. 2017; Pan et al. 2020).

Concerning ariboflavinosis diagnosis, riboflavin status may be evaluated through biological markers, namely, the erythrocyte glutathione reductase activation coefficient (EGRAC). In fact, the antioxidant enzyme, glutathione reductase is FAD dependent and riboflavin is promptly converted into the cofactors, FAD or FMN, following its intestinal absorption and hence the maximum enzymatic activity is solely obtained once the proper riboflavin absorption occurs. The EGRAC determines the enzyme activity prior to and *post* exogenous FAD supplementation in haemolysed erythrocytes (EFSA 2017). The erythrocyte glutathione reductase activation coefficient is determined as the ratio between the *in vitro* glutathione reductase activity in the presence of flavin adenine dinucleotide (FAD) to that without added FAD (Gibson 2005; Rivlin 2010). An EGRAC value of 1.2 or lower is generally utilized to denote an adequate riboflavin status, values within the interval 1.2-1.4 are indicative of a borderline deficiency, whilst values higher than 1.4 indicate riboflavin deficit (Gibson et al. 2005; Rivlin 2010). Nonetheless, it should be pointed out that a higher EGRAC value may not be indubitably interrelated with the level of riboflavin deficiency. Additionally, the EGRAC value cannot be used in individuals having a glucose-6-phosphate deficiency (McCormick 2010).

Ariboflavinosis is more prevalent in underdeveloped countries, consequence of the dietary composition, which comprises only small portions of dairy products and meat (Ashoori and Saedisomeolia 2014), hence not providing proper amounts of riboflavin intake and, concerning vitamin deficiencies, riboflavin hypovitaminosis is the most prevailing (Mosegaard et al. 2020). From a clinical standpoint, riboflavin deficiency symptomatic manifestations are rarely documented in developed countries, albeit the subclinical stage of the deficit, detectable by an alteration in the biochemical indices, might be found in a pronounced percentage of the global population (Mosegaard et al. 2020; Xin et al. 2017). Some developed countries (e.g., United Kingdom and United States) have been reported to present increased prevalence of low serum riboflavin (Ashoori and Saedisomeolia 2014). In a survey conducted in the United Kingdom, children and young adults (particularly 15-18 year-old girls) revealed deficient riboflavin status, which increased with age. The riboflavin status was directly associated with insufficient dietary intake, which in turn was correlated with milk consumption (Mosegaard et al. 2020; Powers 2003).

Riboflavin deficiency has also been demonstrated in the elderly population. It has been reported that in the United Kingdom and the United States intake was insufficient in 10-41% of the elderly (Mosegaard et al. 2020), being that in the United States deficiency was between 10 and 27%. Nonetheless, in the United Kingdom, a disparity was observed between deficiency assessed by dietary intake or by EGRAC. While dietary intake pointed to values lower than 10%, EGRAC registered values around 40%. It was hypothesized that that the discrepancy could be the result of the increased intake necessity amongst the elderly consequence of inefficient absorption (Powers 2003; Mosegaard et al. 2020).

The changes in lifestyle patterns, namely, alterations in diet, with lower (or even devoid, in the cases of vegetarians and vegans) meat, eggs and dairy products consumption, have been shown to decrease the riboflavin daily intake. Increased physical activity may lead to the vitamin deterioration. Either situation results in augmented risk of riboflavin deficiency (Mosegaard et al. 2020).

One additional important factor is the fact that 10-15% of the world population has limited capacity to absorb and utilize riboflavin, which is inherited, resulting in a possible worldwide riboflavin deficiency. Several countries in Europe registered 7-20% riboflavin-deficient individuals and in the United Kingdom, EGRAC revealed that more than half of non-elderly adults (54%) possessed, at least, marginal deficiency (Marashly and Bohlega 2017).

Riboflavin supplementation appears to be a suitable approach for the reversion of deficiency in most of the pathologies (Mosegaard et al. 2020). Indeed, cereal based products have been fortified with vitamins since the beginning of 1940 as a response to the health problems associated to hypovitaminosis affecting a prominent percentage of the population. Owing to the degradation of the vitamins elicited by processing conditions (e.g., milling of grains, which may remove 65-80%), programs focusing on fortification of flours with riboflavin, thiamine and niacin were developed in the United States, as well as in some European countries, which continue to date (WHO/FAO 2006). In Canada, riboflavin enrichment of flour is mandatory since 1976 (Nathoo, Holmes, and Ostry 2005), and in 2019, there were 62 countries with fortification standards concerning riboflavin (Titcomb and Tanumihardjo 2019). In the developed countries fortification of flours and cereals contributes to accomplish the recommended intake levels of these vitamins (WHO/FAO 2006), and throughout the world vitamin B2 has been exogenously supplemented in a wide variety of products, which comprise margarine, wheat flour, pasta, corn masa flour, pre-cooked maize flour and maize flour (WHO/FAO 2006).

1.10. Riboflavin Biosynthesizes by Lactic Acid Bacteria

One of the first documented LAB biosynthesizing riboflavin was *Lc. mesenteroides*, which was isolated from Idli, a traditional fermented Indian rice cake (Rajalakshmi and Vanaja 1967).

In the past decades, novel riboflavin producing LAB have been isolated from a multitude of ecological niches, namely the human intestine, dairy products, and their potential concerning food biofortification applications has been highlighted.

Bhushan and colleagues (2020) used fecal matter and fermented milk products from Indian villages to explore the existence of riboflavin producing strains. The isolation resulted in 68 lactobacilli strains, four of which were determined to be riboflavin-producing strains and were designated BBC32A, BBC32B, BBC33 and BIF43 (the first three isolated from buffalo milk and the latter from feces from an one year-old infant). The bioavailability of vitamin B2 produced by the isolates was confirmed. The bacterial strain presenting the highest production was BBC32B and the lowest was BIF43 (319 and 257 μg L^{-1}, respectively). The genomic DNA was screened for the presence of *rib*G, *rib*B, *rib*A and *rib*H genes involved in vitamin B2 production, revealing that all four strains possessed these genes. These isolates were identified as *L. plantarum* analyzing the 16S-23S rRNA intergenic spacer region and 23S rRNA flanking areas.

In a study performed in China, Ge and colleagues (2020) isolated LAB from pickle-juices produced with different vegetables, finding 90 riboflavin-producing strains with vitamin concentrations ranging from 0.096 to 0.734 mg L^{-1}; 27 of these synthetized over 0.5 mg L^{-1}. The three strains responsible for the highest biosynthesis were identified, based in 16S rDNA, as *L. plantarum*. The strain with the highest riboflavin production (0.7 mg L^{-1}) was found to be *L. plantarum* RYG-YYG-9049 (No. 49), which was additionally used for exploitation of the roseoflavin-resistance spontaneous mutagenesis methodology to select riboflavin-overproducing mutants. Ten mutants presenting enhanced vitamin B2 synthesis were isolated, with mutant RYG-YYG-9049-M10 (the highest producer) eliciting a 3-fold increase in riboflavin concentration (2.33 mg

L^{-1}), comparatively to the wild type strain (0.73 mg L^{-1}). The authors hypothesized the discovery of a novel mechanism concerning the improvement of riboflavin biosynthesis by LAB, owing to the finding of a novel roseoflavin-induced mutation type, namely, a gene insertion located within the upstream regulatory region of the riboflavin operon.

Hati and co-workers (2019), also in India, utilized rice-based traditional fermented beverages as the basis to isolate five LAB strains which DNA analysis revealed belonging to three different species, namely, *L. fermentum* (KGL2, KGL4 and WTS4), *L. plantarum* (KGL3A) and *L. rhamnosus* (RNS4). Amongst the five, the strain presenting the best performance concerning riboflavin biosynthesis was found to be *L. plantarum* KGL3A, which was able to produce 0.7 μg mL^{-1}.

A research conducted by Santos-Mendoza and colleagues (2018) reported riboflavin-producing LAB isolated in different phases of the manufacturing process of a Peruvian traditional product, namely, "Tunta" (native potato based). One hundred and nine LAB were isolated. Thirty two of these isolates were vitamin B2 producers: eleven identified as *Lc. mesenteroides*, four of which presenting high producing features (334 to 426 ng mL^{-1}). The remaining isolates were *Lc. mesenteroides* subsp. *dextranicum* (5), *L. curvatus* (12) and *L. sakei* (4).

Another Peruvian traditional product, namely, "tocosh," which consists in fermented potatoes, was also found to be a reservoir of LAB with the ability to produce the vitamin B2 (Jiménez et al. 2018). The study assessed samples at different production times (freshly harvested, one and eight-month production), finding LAB predominance in freshly harvested potatoes. Several strains of *L. sakei* and *Lc. mesenteroides* were unveiled to produce riboflavin, amongst which *Lc. mesenteroides* were the highest producers with 32% of the strains producing above 100 ng mL^{-1}.

South America, and particularly, the Andean region, has been prolific in the discovery of riboflavin-producing LAB strains. In 2017, Carrizo and colleagues (2017) reported the first study in which several strains of bacteria that synthesized vitamin B2 were isolated from amaranth grains (a "pseudocereal" cultivated in South and Central America, Africa, Asia and India) and spontaneous amaranth sourdough fermentation (SASF).

Bacterial species differed according to the matrix, and amaranth grains contained *Leuconostoc* and *Enterococcus* strains, while SASF comprised *L. plantarum* and *L. rhamnosus*, except one strain, which was identified as *Enterococcus mundtii* CRL2094. The highest riboflavin producers were, in grains, *Lc. mesenteroides* subsp. *mesenteroides* CRL2131 and *E. durans* CRL2122, originating over 250 ng mL^{-1}, and in SASF, *L. plantarum* CRL2107, which produced 158 ng mL^{-1}.

Dairy products constitute a well-recognized source of LAB. Zago and colleagues (2017) isolated three strains of *L. plantarum* (Lp790, Lp813 and Lp998) from Italian and Argentinean cheeses, which were assessed for riboflavin biosynthesis. Although the authors documented that all three strains possessed the genes involved in the vitamin B2 production, only Lp998 was found to synthesize a modest amount of 6.36 ng mL^{-1}.

Quinoa grains (QG) and quinoa spontaneous sourdough (QSS) have also been exploited concerning LAB isolation (Carrizo et al. 2016). Genetic typing enabled the identification of forty-four strains, belonging to four genera, namely, *Lactobacillus*, *Pediococcus*, *Leuconostoc* and *Enterococcus*. Spontaneous sourdough presented higher bacterial diversity than QG. Although *Enterococcus gallinarium* was the prevailing species in the three quinoa grains tested, the prevalent species comprised in quinoa sourdough varied according to quinoa flours used to produce QSS. Concerning riboflavin biosynthesis, LAB isolated from QSS produced the highest vitamin concentrations, ranging between 6 and 360 ng mL^{-1}, and *L. rhamnosus* was the highest producing species (over 270 ng mL^{-1}). With respect to the strains isolated from QG, the riboflavin production ranged between 13-20 ng mL^{-1}.

In an attempt to exploit the potential capability of indigenous LAB of goat milk to biosynthesize riboflavin *in situ*, as a more economically appealing approach than chemical fortification, Pacheco Da Silva and colleagues (2016) documented the first LAB isolation from goat milk and cheese. The authors stated that seven isolates produced vitamin B2 in concentrations above the average, mainly extracellularly secreted, namely *Streptococcus thermophilus*, *Lc. lactis* subsp. *lactis*, *Streptococcus lutetiensis* and *Streptococcus infantarius* (total content of 532, 523, 508 ng mL^{-1} and 437 ng mL^{-1}, respectively).

Li and colleagues (2016) isolated *L. plantarum* LZ227 from raw cow milk, presenting the genetic composition required for riboflavin biosynthesis, and the production was confirmed through a turbidimetric assay (growth assessed by absorbance), using an auxotroph strain.

In a study conducted by Cárdenas and co-workers (2015), the authors demonstrated that *L. fermentum* CECT 5716, previously isolated from human breast milk (Rocío Martín et al. 2003), produced intracellular and extracellular riboflavin in concentrations of 290 and 333 ng mL^{-1}, respectively.

Russo and colleagues (2014) isolated, from an Italian variety of durum wheat, several riboflavin producers, namely, *L. plantarum* (15), *Lc. mesenteroides* (3) and *L. fermentum* (1). The wild-type strain *L. fermentum* PBCC11, isolated from sourdough, was subjected to a roseoflavin-resistance assay to select mutants with the ability to overproduce riboflavin. Nonetheless, amongst the resulting fifteen isolates, only seven over-produced to the desired amount (above 1 mg L^{-1}) and, unexpectedly, three strains were not overproducers. Among the 12 overproducers, mutations in *rib* operon were observed at positions 240 (A to G) and 236 (T to A) in the highest and lowest vitamin B2 producers, respectively, synthetizing riboflavin above 1 mg L^{-1}. *L. fermentum* PBCC11.5 was the highest producer (1.2 mg L^{-1}). The three mutants that were able to resist roseoflavin exposure, and not being identified as riboflavin overproducers, did not harbor any mutation in the riboflavin operon.

In India, where several research works have been performed focusing this subject, Guru and Viswanathan (2013) isolated six vitamin B2-producing LAB from curd and cheese samples from households. The strains were identified as *Lactobacillus acidophilus* and *Lc. lactis*, and riboflavin biosynthesis assays revealed that *L. acidophilus* yielded higher content than *Lc. lactis* (2930 *versus* 2610 μg L^{-1}, respectively).

The cereals were explored by Turpin and colleagues (2011), which used pearl millet slurries to isolate LAB identified as belonging to *Pediococcus* and *Lactobacillus* genera, with *L. fermentum* being the dominating species. The genetic analysis disclosed that all isolates possessed most of the genes pivotal for riboflavin biosynthesis, which allowed to consider them putative producers.

Two *L. plantarum* strains, namely, *L. plantarum* UNIFG1 and UNIFG2, were isolated from Italian artisanal natural sourdoughs and were found to present riboflavin producing capability. The roseoflavin-resistant screening enabled the isolation of seventeen mutants, producing riboflavin concentrations of 488-642 μg L^{-1} (Capozzi et al. 2011).

Jayashree and co-workers (2010) found that, out of thirty-three LAB isolated from fermented milks from households in rural India, twenty four were riboflavin producers, and the isolate with the best biosynthetic capability was identified as *L. fermentum*, which produced a vitamin B2 content of 2.29 mg L^{-1}.

An innovative method was explored by Chen and co-workers (2017) to screen for strains of *Lc. lactis* with ability to synthesize riboflavin. The methodology consisted in the employment of microfluidic droplet technology to perform high-throughput screening. By using fluorescence-activated droplet sorting, it was possible to isolate mutants with high riboflavin production, from a mutagenesis library.

1.11. Riboflavin Food Sources

Humans (*Homo sapiens sapiens*), as previously mentioned, are incapable to biosynthesize riboflavin and hence must acquire this essential vitamin from exogenous sources (Fitzpatrick et al. 2012). Indeed, the main sources of the aforementioned vitamin are milk, dairy and meat products (particularly offal). Likewise, other important sources of riboflavin comprise cereals, meats and fatty fish, egg white, along with specific fruits, whole grains and dark green, leafy vegetables (Powers 2003).

Several studies and reports have been dedicated to assess the vitamin B2 levels in populations, as well as to which food products consumption contribute to the observed values. A recently published research work (Auclair, Han, and Burgos 2019) analysed the contribution of milk and dairy-free alternatives, namely plant-based beverages (soy, coconut and almond), to the nutrient intakes in Canadian adults, demonstrating that 24.4% of the riboflavin intake occurred by consumption of these products, with milk being the primary source. The Spanish ANIBES (Anthropometric data, macronutrients and micronutrients intake, practice of physical activity, socioeconomic data and lifestyles in Spain) study (Mielgo-Ayuso et al. 2018) evaluated the dietary intake of B-complex vitamins in individuals aged 9 to 75 years, and reported that 72% of the assessed population fitted the EFSA riboflavin recommended intake levels.

Table 4. Riboflavin content in several food products (Roe et al. 2017)

Product	mg $100g^{-1}$
Sheep liver (fried)	5.65
Pig kidney (fried)	3.70
Liver-type sausage	1.88
Bovine heart (raw)	0.91
Cheese, Roquefort	0.65
Sheep milk	0.36
Cow milk (whole)	0.23
Goat milk	0.15
Crabmeat (cooked)	0.86
Mackerel (baked)	0.41
Sardines (cooked in water)	0.39
Marine shrimps or prawns (cooked)	0.33
Popped cereals	1.30
Non-fermented tea leaves (green or white tea)	0.95
Dried mushrooms	3.47
Dulse (red algae)	0.58

Furthermore, the mentioned study unveiled a mean riboflavin intake of 1.44 mg day^{-1}, which is above the EFSA recommendation of 1.3 average requirement (EFSA 2017), and that the main vitamin B2 sources were milk and dairy products. Kehoe and co-workers (2018) investigated the dietary intake in Ireland, in adult population aged 18 to 90 years, finding that 18% presented deficient riboflavin intakes, while mean daily intake was 2.5 mg day^{-1}. The blood riboflavin content revealed that 61% possessed vitamin B2 levels above the value considered for riboflavin deficiency (EGRAC >

1.3), showing a considerable percentage of riboflavin deficient adults. The main dietary riboflavin sources were milk (20-24%), followed by meat and meat products (14-18%), and ready-to-eat breakfast cereals (6-14%). Tea and alcoholic beverages were also significant contributors (8-10%).

An extensive nutrient composition database has been published by EFSA (Roe et al. 2017), in which the composition of food products from 14 European countries was assessed in terms of several vitamins and minerals contents, among which, riboflavin (vitamin B2). The riboflavin content in some food products is depicted in Table 4. It can be observed that meat products contained the highest vitamin B2 amounts, albeit several other products also comprised high contents, which demonstrates that an adequate diet may provide the necessary intake of riboflavin.

1.12. Development of Riboflavin Biofortified Functional Foods

As previously detailed, one emergent approach to isolate vitamin B2-overproducing LAB strains relies on the exploitation of the intrinsic resistance to roseoflavin. Notably, owing to the fact that the mentioned strains are spontaneous mutants, and hence not considered bioengineering derivatives, the stated roseoflavin-resistant organisms may be intended for the *in situ* vitamin B2 biosynthesis originating riboflavin bio-enriched foods.

Due to the increasingly worldwide reported riboflavin hypovitaminosis (ariboflavinosis), investigators have concentrated their efforts on the discovery and development of natural approaches to improve the vitamin B2 content in foods. In this sense, LAB emerged as appealing microorganisms to formulate novel biofortified fermented products, due to the notable capability to biosynthesize riboflavin *in situ*. Despite the prominent number of vitamin B2-producing LAB identified to date, scarce studies have addressed the *in vivo* impact of the ingestion of biofortified fermented foods towards ariboflavinosis prevention.

Yépez and colleagues (2019) used five *L. plantarum* strains, previously isolated from an Argentinian fermented drink "chicha" (Elizaquível et al.

2015), and naturally fermented potatoes "tocosh," from the Peruvian Andes (Jiménez et al. 2018), to uncover putative riboflavin-overproducing mutants following roseoflavin exposure (Yépez et al. 2019). The authors performed HPLC analysis of the vitamin B2 synthesized by the overproducing-mutants and assessed the technological features of those strains. Concerning the analysis, it emerged the notable potential, and the feasibility, to use *L. plantarum* M5MA1-B2 to formulate riboflavin fortified (enriched) cereal-based fermented beverages (Yépez et al. 2019). The kefir-like cereal-based products were prepared with oat, maize and barley flour (milk was used as control), which were fermented by commercial starter cultures and also co-inoculated with the selected strain. The presence of *L. plantarum* M5MA1-B2 in the fermentation process resulted in higher vitamin B2 contents in all three cereal-based beverages, with the highest concentration (*ca.* 1.6 mg mL^{-1}) being obtained in oat-based kefir-like. In this sense, one 100 g serving would provide 11.4% of the RDA.

Thompson and colleagues (2020) exploited the fermentative potential of vegetables by LAB, in order to enhance the B-vitamins content, particularly, riboflavin. In the mentioned study, cauliflower, white beans, and a mixture of both (50:50) were employed to evaluate the impact, in the riboflavin content, of the fermentation process performed by four strains of *L. plantarum*, three of which were isolates of human origin, namely, strains 299v, 299 and Heal19, and one food isolate, Lp900. The authors highlighted the notable performance of *L. plantarum* Lp900 in fermenting the mixture of cauliflower and white beans, eliciting an increase of 113% in the riboflavin concentration in comparison with the unfermented vegetables. A vitamin concentration of 91.6 μg *per* 100 g of fresh weight was obtained, whereas the lowest value (level) was 75.6 μg 100 g^{-1} of fresh weight, corresponding to an increase of 76% of the initial content of riboflavin. Noteworthy, comparatively to the average requirement (AR) and the population reference intake (PRI) of 1.3 and 1.6 mg *per* day, respectively, recommended by EFSA (EFSA 2017) for adults, the quantity of the vitamin obtained in the fermented product was pronouncedly low and hence appeared not to constitute a viable approach to suppress a potential hypovitaminosis.

Ge and co-workers (2020) reported the use of an overproducing riboflavin strain, *L. plantarum* RYG-YYG-9049-M10, to investigate the fermentation of soymilk. Different parameters (time and temperature) were evaluated to formulate a fortified beverage with a prominent *in situ* biosynthesis of the vitamin B2. The highest concentration (2.92 mg L^{-1}), obtained in the optimal conditions (20 h at 37 °C), represented an increase of *ca.* 10-fold in the riboflavin content in comparison with the unfermented soymilk (0.27 mg L^{-1}), providing in 100 mL of product 21% of the RDA for adults.

Zhu and colleagues (2020) also developed a novel riboflavin-enriched soybean milk fermented by vitamin B2-overproducing lactobacilli strains, documenting an improvement of 19- and 9.5-fold in the riboflavin content following 16 h of the fermentative process performed by *L. fermentum* UFG169 (from 0.2 to 3.8 µg mL^{-1}) and *L. plantarum* UFG10 (from 0.2 to 1.9 µg mL^{-1}), respectively. The authors proposed to evaluate a putative correlation between the *in vitro* antioxidant capacity of the fermented soybean milk and the riboflavin concentration, since scarce information addressing this issue is available. The ABTS and DPPH assays unveiled the achievement of an enhanced free radical scavenging capacity as a consequence of the increase in extracellular vitamin B2 content, corroborating the formulated hypothesis.

Bhushan and co-workers (2020) reported the well-recognized unique potential, concerning riboflavin biosynthesis, amongst phenotypically closely related strains, particularly four *L. plantarum* isolates. The authors hypothesized that repression of a *rib* genes operon triggered by the riboflavin comprised in the unfermented soymilk, putatively strain dependent, may have been the underlying reason for the observed discrepancy in the extracellular content of the vitamin B2. In the mentioned study, the wild-type strain *L. plantarum* BBC32B, isolated from Dahi fermented milk, presented the highest riboflavin biosynthesis, *ca.* 319 µg L^{-1} during the logarithmic phase. The BBC32B isolate was further employed to ferment freshly prepared soymilk, containing a scarce riboflavin content, generating a fermented soy curd comprising a B2-biofortification level of 49.05% (700 μg L^{-1}). The authors highlighted that the riboflavin content of the BBC32B plant-based beverage corresponded to approximately half of the RDA for adults (1.3 mg) and was comprised within the range of the RDA intended for children (0.5 to 0.9 mg).

Carrizo and colleagues (2020) documented the *in vivo* evaluation of the effectiveness of a novel biofortified riboflavin quinoa pasta concerning ariboflavinosis prophylaxis. The authors employed a 2-strain cocktail encompassing two riboflavin producing *L. plantarum* isolates, namely CRL 2107 and CRL 1964, to perform the fermentation of quinoa (*Chenopodium quinoa* Willd) flour in order to formulate a B2-enriched pasta (quinoa gnocchi) comprising a vitamin content of 5.0 ± 0.5 μg g^{-1}. In this sense, an 80 g-portion of the quinoa gnocchi will represent 31% of the recommended daily intake for riboflavin. In the mentioned study, *in vivo* experiments were performed, in which BALB/c mice were fed the bio-enriched fermented product for 42 days, along with a riboflavin-deprived balanced food regimen. The authors argued that the vitamin B2 bioavailability in the fermented quinoa was comparable to that of the synthetic form, based on the demonstrated similar effectiveness in preventing riboflavin deficiency (hypovitaminosis). Furthermore, concerning the vitamin B2 role in the gastrointestinal tract development, the histological evaluation of the small intestine unveiled an unaltered maturation of the organ, with inexistent degeneration in villi structure, a well-recognized feature related to riboflavin nutritional deficit.

Owing to the low content of riboflavin, soymilk has been extensively investigated concerning its biofortification provided by the *in situ* vitamin biosynthesis by LAB, a more economical appealing approach than the chemical vitamin supplementation and in line with the consumers demand for healthier food products.

Levit and co-workers (2017) assessed the anti-inflammatory potential of a soymilk fermented by the riboflavin overproducing strain, *L. plantarum* CRL 2130, in a mouse model of acute colitis elicited by trinitrobenzene sulfonic acid (TNBS). The vitamin B2 content in the fermented soymilk was previously determined to be 1.6 μg mL^{-1}, a 6-fold higher value than the initial riboflavin concentration of the non-fermented beverage (Juarez del Valle et al. 2014). The macroscopic and histological evaluation of the large intestine of mice fed with 3 mL of this bio-enriched beverage *per* day unraveled scarce intestinal damages, presenting unaltered mucosa muscular layer. Moreover, the authors observed that mice administered with the functional beverage exhibited a lower weight loss and a significant decreased microbial translocation towards the liver, in comparison with the animals fed with the unfermented soymilk. Pertaining to the intestinal cytokine profile, the results demonstrated lessened levels of pro-inflammatory cytokines, namely, tumor necrosis factor (TNF)-α, interleukin (IL)-12, IFN-γ, IL-17 and monocyte chemoattractant protein (MCP)-1, pointing out an analogous potential to the chemical vitamin form to revert the gut inflammatory response. This was the first study documenting the *in vivo* positive impact of a riboflavin-biosynthesizing bacterial strain in a murine model of intestinal inflammation.

In a previous work, Juarez del Valle and co-workers (2016), envisaging the putative capability of the soymilk fermented with *L. plantarum* CRL2130 to revert and prevent riboflavin deficiency (ariboflavinosis), conducted *in vivo* bioassays to assess its efficacy. In order to mimic a riboflavin deficiency, a depletion-repletion murine model was employed, in which BALB/c mice were administered with a 21-days regimen deprived of vitamin B2 (riboflavin devoid), followed by an equal period (21 days) during which the animals were fed the riboflavin deficient diet complemented with the B2-biofortified soymilk. Pertaining to the assessment of the beverage prophylactic potential towards ariboflavinosis, mice were administered for 42 successive days a B2-deficient diet supplemented with the riboflavin-bio-enriched soymilk. It is well-recognized a lower body weight increase once experimental animals experience riboflavin privation. Noteworthy, the authors demonstrated that upon the depletion period, the animals fed with the CRL2130-fortified soymilk presented a growth rate similar to those administered with the soymilk supplemented with the commercial form of the vitamin or a control diet (7 mg *per* Kg). With respect to the prophylactic model, the same growth rate pattern was observed between mice provided with the diets comprising the soymilk containing the commercial vitamin and the one biofortified with the vitamin B2 overproducing strain, CRL2130.

The riboflavin status was evaluated through the biological marker, EGRAC. Upon the depletion phase, EGRAC of mice fed with the biofortified CRL2130-fermented soymilk declined to values below 1.3, similarly to those administered with the chemical form supplemented beverage. The same normalized EGRAC values were observed concerning the prophylactic approach. The authors further assessed the intestinal villi height as an indicator of the duodenum development and found that in both evaluated approaches, the vitamin B2, whether administered in the biofortified or chemical supplementation, elicited maintenance (ariboflavinosis prevention) and reversion of villi size/morphology, being effective in preventing hypertrophy. The authors concluded that the B2-bio-enriched beverage may be a feasible riboflavin source concerning individuals with clinical and subclinical ariboflavinosis. It was pointed out that a 200 mL-dose of the CRL2130-fermented soymilk would represent a B2 intake of 0.37 mg, hence constituting 28% of the recommended average daily intake.

Kaprasob and colleagues (2018) investigated the notable capability of five probiotic strains to ferment the underrecognized industrial by-product cashew apple juice (CAJ), towards the enhancement of its vitamin B2 content. Upon 48 h of fermentation, it was observed that the most prominent *in situ* riboflavin biosynthesis were provided by *Lactobacillus casei* TISTR 390 (137.15 mg L^{-1}) and *L. acidophilus* TISTR 1338 (131.54 mg L^{-1}). The authors noticed a decrease in the riboflavin concentration of the fermented CAJ throughout the initial 24 h, owing to the vitamin B2 metabolic requirements of LAB, or the putative conversion of riboflavin into the cofactors, FAD and FMN. Noteworthy, fermented CAJ may be an innovative vitamin B2-biofortified juice.

Oguro and co-workers (2017) reported the B2-bioenrichment of a traditional Japanese beverage, *koji* amazake, following the fermentative process performed by a three-strain *L. sakei* UONUMA cocktail (previously isolated from snow caverns), which also conferred a renewed flavor. The authors observed a significant improvement in the vitamin B2 content of *ca.* 1.45-fold (748.6±42.9 μg 100 mL^{-1}) of the novel fermented functional amazake, in comparison with the unfermented beverage (517.4± 64.9 μg 100 mL^{-1}).

A different food matrix from the commonly explored in riboflavin fortification was used by Russo and co-workers (2015), which investigated the viability of using vitamin B2 overproducing LAB, namely, *L. plantarum* B2 and *L. fermentum* PBCC11.5 (isolated from sourdough, (Russo et al. 2014)), to increase the riboflavin concentration in fresh-cut cantaloupe. The fresh-cut, peeled, fruit pieces were dipped into a buffer solution containing each of the bacterial strains, independently, after which the fruit pieces were air dried. In addition to the determination of the riboflavin content originated by the exogenous supply of the strains, samples were also stored at 4 °C and the study assessed the vitamin variation throughout 11 days of storage. The assessment of the riboflavin contents was performed at 0, 4, 8 and 11 days, and revealed that the initial vitamin concentration of 6.5 μg 100 g^{-1}, present in non-inoculated melon, decreased throughout storage, with a final content 15% lower. Regarding inoculated samples, behavior was the opposite, with *L. plantarum* B2 samples increasing 33% riboflavin content during the first four days, reaching a 2-fold increase by the end of the storage period, registering 12 μg 100 g^{-1}. Samples inoculated with *L. fermentum* PBCC11.5 presented, at 11 days of storage, an increase in vitamin B2 level to 10 μg 100 g^{-1}.

In a research work performed by Russo and colleagues (2014), a strain of *L. fermentum* (*L. fermentum* PBCC11), isolated from sourdough, was subjected to roseoflavin-resistance assay to select mutants with the ability to overproduce riboflavin. Nonetheless, of the resulting fifteen isolates, only seven over-produced to the desired extent (> 1 mg L^{-1}) and, unexpectedly, three did not comprise this phenotypic feature. Among the 12 overproducing bacteria, mutations in the *rib* operon were observed at positions 240 (A to G) and 236 (T to A) in the highest and lowest producers, respectively. The three mutants that were able to resist roseoflavin exposure, albeit not having the ability to overexpress the riboflavin operon, did not harbor mutations in *rib* genes. The highest producer was found to be *L. fermentum* PBCC11.5, which was further used in the fermentation process of bread, using the parental (wild-type) strain as control. The resulting vitamin B2-enriched bread possessed about double the riboflavin content than that of the bread fermented with the wild-type strain (7 ug g^{-1} *versus* 3.3 ug g^{-1}). In this sense, the riboflavin content of a 100 g portion of the fortified bread would be able to provide 0.66 mg, which represents, concerning adults, 50% of the RDA.

Capozzi and co-workers (2011) isolated, from durum wheat flour, a multitude of presumptive LAB (60), which were screened by PCR for the *rib* operon, and tested regarding roseoflavin-resistance, revealing two *L. plantarum* strains UNIFG1 and 2 with riboflavin-overproducing capability. The bacterial strains were thereafter used in the process of bread fermentation and in a pre-fermentation stage in pasta manufacturing, with the goal of increasing the products' riboflavin contents. A cocktail encompassing the two strains was employed in both experiments in bread and pasta, and *L. plantarum* WCFS1 was utilized as a control strain since, due to possessing an incomplete *rib* operon, it was not able to biosynthesize riboflavin. The improvement in vitamin B2 was successful, since in bread its content increased 3-fold (from 0.24 mg $100g^{-1}$ to 0.68 mg $100g^{-1}$), and nearly doubled in pasta (0.45 μg g^{-1} to 2.48 μg g^{-1}). Capozzi and colleagues (2011) hypothesized that the manufacturing process may also impact the riboflavin content, having the authors argued that the increase observed following extrusion may have resulted from bacterial lysis due to the high pressures applied in that specific process. It was also pointed out the importance of nutrient availability, as higher vitamin B2 contents were observed in remilled semolina, which presented higher amount of damaged starch, eliciting an increased availability of soluble sugars to be metabolized by LAB.

The enhancement of the nutritional features of food products, riboflavin included, is not a recent subject, as may be observed in the research work published by Camacho and colleagues (1991), in which several species of *Lactobacillus* (*L. acidophilus*, *L. buchneri*, *L. cellobiosus* and *L. fermentum*) were used with the objective of enhancing the nutritional quality of lupine through fermentation. The strains presented different behaviors, each modulating the composition of specific compounds, and *L. acidophilus* B-1910 was found to be related to increased riboflavin content. The study elucidated that the origin of vitamin B2 relied on the bacterial biosynthesis, and was not attributed to the phytate complex solubilization, being the latter hypothesis discarded since the observed decrease in phytate content was moderate. The utilization of *L. acidophilus* B-1910 in the fermentation of the lupine beans, allowed the authors to achieve a lupine beverage with improved riboflavin content.

The previously detailed biofortified fermented foods, along with the corresponding vitamin B2 content are outlined in Table 5.

Table 5. Riboflavin content in LAB fermented food products

Foods	Microorganism	Vitamin B2 content	Reference
Rice-based traditional fermented beverages	*Lb. plantarum* KGL3A	0.7 μg mL^{-1}	Hati et al. 2019
Tunta	*Leuconostoc mesenteroides*	334 to 426 ng mL^{-1}	Santos-Mendoza et al. 2018
Tocosh	*Leuconostoc mesenteroides*	100 ng mL^{-1}	Jiménez et al. 2018
Amaranth grains	*Lc. mesenteroides* subsp. *mesenteroides* CRL2131 *E. durans* CRL2122	250 ng mL^{-1}	Carrizo et al. 2017
Spontaneous amaranth sourdough fermentation (SASF)	*Lb. plantarum* CRL2107	158 ng mL^{-1}	Carrizo et al. 2017
Quinoa spontaneous sourdough (QSS)	*Lb rhamnosus*	270 ng mL^{-1}	Carrizo et al. 2016
Goat milk	*Streptococcus thermophilus* FP34	532 ng mL^{-1}	Pacheco Da Silva et al. 2016
Curd and cheese	*Lb. acidophilus* 5	2930 μg L^{-1}	Guru and Viswanathan 2013
Curd and cheese	*Lc. lactis*	2610 μg L^{-1}	Guru and Viswanathan 2013
Fermented milks	*Lb. fermentum*	2.29 mg L^{-1}	Jayashree et al. 2010
Oat-based kefir-like beverage	*Lb. plantarum* M5MA1-B2	11.4% of RDA *per* 100g	Yépez et al. 2019

Table 5. (Continued)

Biofortified Foods	Microorganism	Vitamin B2 content	Reference
Cauliflower and white beans	*Lb. plantarum* Lp900	91.6 μg *per* 100g (fresh weight)	Thompson et al. 2020
Soymilk	*Lb. plantarum* RYG-YYG-9049-M10	2.92 mg L^{-1}	Ge et al. 2020
Soybean milk	*Lactobacillus fermentum* UFG169	3.8 μg mL^{-1}	Zhu et al. 2020
Soy curd	*Lb. plantarum* BBC32B	700 μg L^{-1}	Bhushan et al. 2020
Quinoa gnocchi	Mix of *Lb. plantarum* CRL 2107 and CRL 1964	5.0 μg g^{-1}	Carrizo et al. 2020
Soymilk	*Lb. plantarum* CRL 2130	1.6 μg mL^{-1}	Juarez del Valle et al. 2016; Levit et al. 2017
Cashew apple juice	*Lactobacillus casei* TISTR 390	137.15 mg L^{-1}	Kaprasob et al. 2018
Koji amazake (traditional Japanese beverage)	Three-strain *Lb. sakei* UONUMA cocktail	748.6 μg 100 mL^{-1}	Oguro et al. 2017
Fresh-cut cantaloupe	*Lb. fermentum* PBCC11.5	10 μg 100 g^{-1}	Russo et al. 2015
Bread	*Lb. fermentum* PBCC11.5	7 μg g^{-1}	Russo et al. 2014
Bread	*L. plantarum* UNIFG1 and UNIFG2	0.7 mg $100g^{-1}$	Capozzi et al. 2011
Pasta		2.48 μg g^{-1}	

1.13. Riboflavin Degradation and Stability in Foods

Riboflavin stability in food products is important since the vitamin degradation may decrease the content present in the product at the time of consumption, and it may also affect the stability of other compounds, impacting the overall quality of the product.

Vitamin B2 heat-stability is widely recognized and riboflavin is considered to be, overall, one of the most stable vitamins, being able to withstand conditions which are detrimental to other vitamins (Burch 2011; Berry Ottaway 2010; EFSA 2017). Riboflavin degradation in food products occurs mainly throughout storage and the principal factor affecting the vitamin stability is the light oxidation (photosensitization).

Nonetheless, other factors, namely water activity, oxygen and metal sulphates, may also present a detrimental impact in stability, although to a lesser extent (Choe, Huang, and Min 2005; Burch 2011).

As mentioned, photosensitization is the main responsible for the riboflavin degradation. Riboflavin is a photosensitizer which, when exposed to light, is cleaved to ribitol and lumiflavin, the latter also being an efficient photosensitizer. Following light exposure, singlet (^{1}Rib*) and triplet-excited (^{3}Rib*) states of riboflavin are generated, being that the singlet is short-lived while the triplet is relatively long-lived, and the resulting triplet may be quenched by other compounds, which consequently leads to their oxidative degradation (photosensitized by riboflavin triplet). Triplet riboflavin is quenched by most amino acids in anaerobiosis, while in the presence of oxygen (which also quenches ^{3}Rib*), tryptophan and tyrosine are the only ones competing to quench. Polyunsaturated lipids, phenols and some vitamins (e.g., folate) also quench ^{3}Rib* and suffer photooxidation sensitized by the triplet. Nonetheless, some molecules present protection against the detrimental effects caused by the riboflavin triplet. Several purine and pyrimidine bases are able to deactivate ^{3}Rib*, thus protecting degradation of other molecules, and carotenoids are known to prevent formation of ^{3}Rib* by inner-filter effects, thus impeding the riboflavin photosensitized oxidation (Cardoso, Libardi, and Skibsted 2012).

These reactions should be considered when products are being developed as, consequence of the reactions previously described, the presence of some compounds may have a detrimental impact, leading to increased light-sensitivity, while others may possess a protective effect. Packaging must also take this aspect into consideration and be performed with materials which minimize, or abolish, light exposure of the products (Mestdagh et al. 2005). Those containers include brown glass, opaque plastic and cardboard containers, among others (Cardoso, Libardi, and Skibsted 2012; Pinto and Zempleni 2016; Mestdagh et al. 2005).

The products with extended shelf-life, for instance butter, cheese and dairy spreads must be particularly considered, as well as products that may encompass in their composition compounds that are susceptible to photosensitized oxidation, namely polyunsaturated fatty acids, as it was previously mentioned (Cardoso, Libardi, and Skibsted 2012).

An example of the riboflavin degradation was the reported loss of 20-80% of vitamin content in milk, as a consequence of exposure to light within 2 hours (Berry Ottaway 2010). The author also stated that light intensity, along with temperature and exposure surface area contributed to the extent of the degradation. However, despite the sensitivity observed in liquid medium (milk), the riboflavin stability may be maintained in other matrices exposed to light, for instance in white bread stored in transparent packages.

The vitamin stability was also mentioned by Saade and Arijaje (2020) regarding breakfast cereals produced with riboflavin enriched white flour, in which the storage at room temperature and low water activity resulted in considerable stability of the riboflavin and, consequently, maintenance of the vitamin content.

All the abovementioned reactions of vitamin B2 with food components and environmental conditions are also valid when riboflavin fortification is concerned and, therefore, must be taken into consideration in the product development.

1.14. Conclusion

Riboflavin (vitamin B2) is a pivotal component of a myriad of cellular metabolic pathways, owing to the fact that this vitamin is the metabolic precursor of the flavin coenzymes FAD and FMN.

Currently, industrial production of riboflavin has transitioned from the chemical synthesis to be solely performed through the fermentative process, owing to the higher economic and ecological suitability. This shifting pattern also occurred due to the remarkable progresses of genetic and bioprocesses engineering. Indeed, the two major fermentation processes are accomplished through the employment of the bioengineered derivatives of the Gram-positive bacterium *B. subtilis* and filamentous hemiacomycete *A. gossypii*.

Vitamin B2 constitutes an essential dietary component, having a critical role in human health and hence ariboflavinosis may elicit a multitude of pathologies and clinical symptoms. Owing to the humans' incapability to synthesize riboflavin, the vitamin requirements must be fulfilled through the dietary intake. Due to inadequate diet, deficient dietary options or malabsorption, this is commonly not achieved, resulting in a considerable percentage of the population presenting, at least, marginal deficiency.

In an attempt to circumvent the prominent worldwide diagnosed ariboflavinosis, investigators have concentrated their focus on the exploitation of natural approaches to improve the vitamin B2 content in foods. The notable potential of LAB to *de novo* biosynthesize riboflavin has been explored and a plethora of novel biofortified fermented foods have been developed. These functional products may constitute a feasible prophylactic, or even therapeutic, approach concerning riboflavin deficiency.

Albeit the significant number of riboflavin-producing LAB identified to date, scarce studies have addressed *in vivo* the impact of the consumption of biofortified fermented food products towards ariboflavinosis reversion or prevention.

2. Vitamin B9

2.1. Background

Folic acid, derived from the Latin word *folium*, that means leaf, was firstly isolated from spinach in 1941. Two years later, in 1943, the pure crystalline form of the compound was synthetized by Bob Stokstad and in 1945 by Angier and colleagues (1945). Pteroylglutamic acid was the name given to folic acid after knowing that it was composed of a pteridine ring, para-aminobenzoic acid and glutamic acid. Later, differences between natural folates and chemically synthesised folic acid were observed. In 1979, Bob Stokstad and colleagues isolated, purified and characterized for the first time many enzymes involved in folate metabolism in mammalian (Hoffbrand and Weir 2001).

2.2. Chemical Structure and Properties

Folate or folic acid, also known as vitamin B9, is an essential vitamin constituted by tetrahydrofolate and its derivatives. As all B group vitamins, vitamin B9 is hydrophilic. Folate, the polyglutamate form, is argued by many authors to be the natural occurring form of the vitamin.

Folic acid is the monoglutamate form of the vitamin, a fully oxidized chemical compound. It is generally chemically synthesized and it is used for supplementation and food and feed fortification. Some authors argue that, although rarely, it can be found in nature.

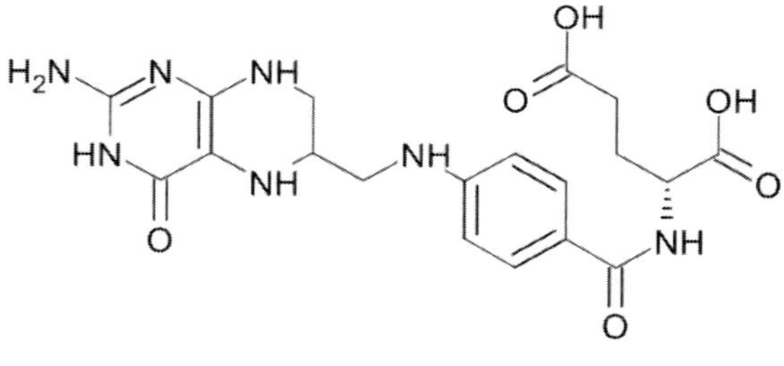

Figure 4. Tetrahydrofolate chemical structure (designed using ACD/ChemSketch; ACD/Labs 2019.2.1).

Natural folates and folic acid are distinct, possessing three different features. Natural folates, in opposition to folic acid, present more than one glutamate residue, reduction to di- or tetrahydroforms, and more single carbon units. Folic acid is more stable than natural folates, and natural folate is less stable than its derivatives. In solution, the less stable forms of the vitamin, reduced forms, are sensitive to oxygen, light and extreme pH values (Zempleni et al. 2013), while the most stable form detains the capability to maintain its chemical structure for 10 h at 100°C, in the dark and at a pH values between 5.0 and 12.0 (Ball 2005).

2.3. Biosynthetic Pathway

2.3.1. Vitamin B9 De Novo Biosynthetic Pathway

Bacteria, fungi and plants have the ability to biosynthesize folate through the *de novo* pathway (De Crécy-Lagard et al. 2007). The synthesis by bacteria has already been described in LAB, namely *L. plantarum* (Rodrigo-Torres et al. 2019) and *Lc. lactis* (Sybesma 2003).

The genes involved in the bacterial biosynthetic pathway of folate are *fol*E, *fol*Q, *fol*B, *fol*K, *fol*P, *fol*C and *fol*A, encoding for GTP cyclohydrolase I (EC 3.5.4.16), dihydropterin triphosphate pyrophosphohydrolase (EC 3.6.1.67), dihydroneopterin aldolase (EC 4.1.2.25), 2-amino-4-hydroxy-6-hydroxymethyldihydropteridine pyrophosphokinase (EC 2.7.6.3), dihydropteroate synthase (EC 2.5.1.15), bifunctional protein synthase/polyglutamil folate synthase (EC 6.3.2.12/17) and dihydrofolate reductase (EC 1.5.1.3), respectively.

A five-membered imidazole ring of guanosine-5'-triphosphate (GTP) is converted into a six-membered dihydropyrazine ring, 7,8-dihydroneopterin triphosphate, in a reaction catalyzed by the enzyme FolE. This step is followed by the conversion of the compound to its monophosphate form, mediated by FolQ, encompassing the removal of phosphate residues. The generated molecule, 7,8-dihydroneopterin is thereafter converted into 6-hydroxymethil-7,8-dihydropterin catalyzed by FolB comprising the concomitant release of glycolaldehyde. The resulting compound is pyrophosphorylated by FolK into 6-hydroxymethil-7,8-dihydropterin pyrophosphate, which in turn is coupled with 4-aminobenzoic acid, catalyzed by FolP, leading to the synthesis of 7,8-dihydropteroate. In a subsequent reaction, the latter compound is converted into dihydrofolate, mediated by FolC, which is further reduced to tetrahydrofolate by FolA. The addition of glutamate catalyzed by FolC leads to the transformation of the compound into tetrahydrofolate polyglutamate (Sybesma 2003; Snauwaert et al. 2015).

The complete pathway is not conserved, presenting some variations between strains. In the experiment conducted by Rodrigo-Torres and colleagues (2019), all the vitamin B9-producing *L. plantarum* strains analysed lacked *fol*Q gene, encoding dihydropterin triphosphate pyrophosphohydrolase, in spite of producing folate. The research team found that all genomes comprised a gene encoding a dITP/XTP pyrophosphatase which explained the missing *fol*Q and proved the complete biosynthetic pathway and production of folate. *Lc. lactis* presented a gene, designated *fol*KE, encoding the bifunctional enzyme, comprising 2-amino-4-hydroxy-6-hydroxymethyldihydropteridine pyrophosphokinase and GTP cyclohydrolase I, contrasting with the previously described, that two different genes encoded the mentioned enzymes (Sybesma et al. 2003). The vitamin B9 *de novo* biosynthetic pathway is represented in Figure 5.

2.3.2. Biosynthesis Regulation

Folates transport and biosynthesis are, in part, regulated by riboswitches (see section 1.3.2) that are associated with mRNA encoding genes, *fol*T (transport gene), *fol*C and *fol*E (biosynthesis genes). Folate riboswitch only binds the active forms, leading to a rejection of the synthetic form, folic acid (Trausch et al. 2011).

2.4. Industrial Production of Vitamin B9

Industrial production of folic acid, the chemically synthetized form, is more interesting than the fermentative biosynthesis of natural folates, since it is more economic and higher yields are obtained.

Synthesis of folic acid results from the reaction of 2,4,5-triamino-6-hydroxypyrimidine, a three-carbon compound of variable structure, and *N*-(*p*-aminobenzoyl)-L-glutamic acid. The reaction occurs with a reducing agent, sodium sulfite, in a water-based mixture.

Synthesis of acid folic derivatives, (6*S*)-methylTHF and (6*S*)-5-formylTHF, is slightly more complex (Rossi et al. 2016).

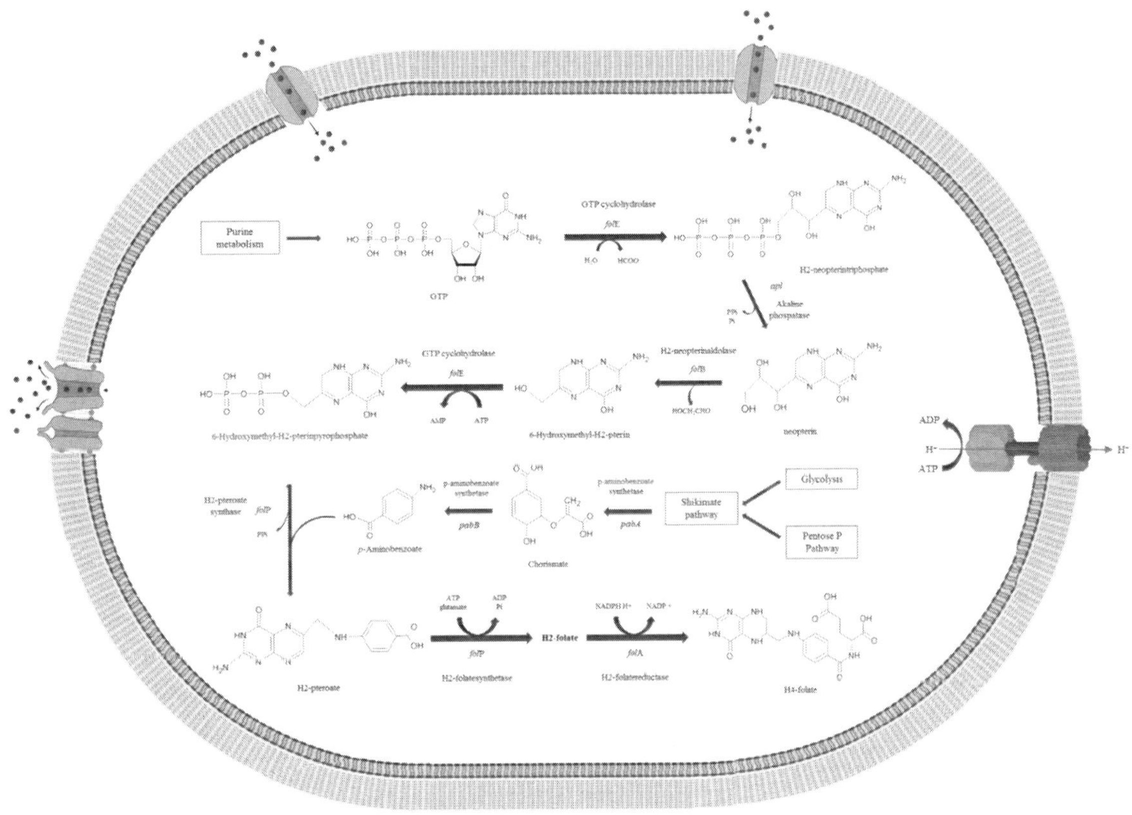

Figure 5. Folate *de novo* biosynthetic pathway (designed using ACD/ChemSketch; ACD/Labs 2019.2.1; adapted from Sybesma (2003)).

2.5. Metabolic Engineering

Microbial production of folates is relatively low to attend the human needs and the concentration of this vitamin in functional foods developed thus far is not sufficient to reach the recommended daily intake and so, the need of increasing this vitamin production is high. Different techniques may be used to increase production, which may be achieved through the generation of random mutants (LeBlanc et al. 2020), and genetic engineering. Genetic engineering contemplates introduction of plasmids, introduction with subsequent elimination of a recombinant plasmid, overexpression, deletion or underexpression of genes and insertion of DNA. Through genetic engineering, Sybesma and colleagues (2003) were able to increase extracellular folate by 8-fold, from 10 to 80 µg L^{-1}, in *Lc. lactis* by overexpressing *fol*KE, while an increase of almost 3-fold was achieved for total folate. Wegkamp and co-workers (2007) increased folate content by overexpressing folate gene and *p*ABA in *Lc. lactis*, reaching 2300 µg L^{-1}, which would not be obtained by overexpressing those genes separately. Nisin induction of *pab*A-*pab*BC independently of the addition or absence of *p*ABA led to a concentration of 2700 µg L^{-1}. Serrano-Amatriain and colleagues (2016) tested different genetic alterations of *A. gossypii*. A 146-fold increase in folate production, reaching 6595 µg L^{-1}, was achieved with a combination of overexpressing three *fol* genes, P_{GPD1}-*FOL2-3,* deleting *AgMET7* and *AgADE12* and underexpressing *AgRIB1* (Serrano-Amatriain et al. 2016).

2.6. Folate Producers

Some plants and microorganisms are able to synthesize this vitamin. There is a considerable number of bacteria and fungi described in the literature as folate producing organisms. These are mainly LAB, some of which isolated from foods like radish, peas, milk and dairy products.

LAB are included in the Qualified Presumption of Safety (QPS) list by Food and Drug Administration (FDA) and European Food Safety

Authority (EFSA) (Rodrigo-Torres et al. 2019; Thompson et al. 2020), having a beneficial use in food fermentations. Some examples of folate producing LAB comprise *Bifidobacterium* spp. (Tamene et al. 2019), *Enterococcus faecium*, *E. lactis*, *E. mundtii* (Salvucci, LeBlanc, and Pérez 2016), *Lc. lactis* subsp. *lactis*, *Lc. lactis* subsp. *cremoris* (Sybesma 2003), *L. plantarum* (Carrizo et al. 2020; Albano, Silvetti, and Brasca 2020; Masuda et al. 2012; Tamene et al. 2019), *L. rossieae* (De Angelis et al. 2014), *L. coryniformis* (Masuda et al. 2012), *L. casei*, *L. paracasei* subsp. *paracasei*, *L. rhamnosus* (Albano, Silvetti, and Brasca 2020), *Lactobacillus delbrueckii* subsp. *bulgaricus*, *Lactobacillus sakei* (Kim et al. 2020; Masuda et al. 2012), *Lactobacillus pentosus* (Salvucci, LeBlanc, and Pérez 2016), *Lactobacillus reuteri* (Kariluoto et al. 2014), *L. fermentum*, *Leucosnotoc* spp., *Pediococcus acidilactici* (Salvucci, LeBlanc, and Pérez 2016) and *Streptococcus thermophilus* (Levit et al. 2019; Kariluoto et al. 2014). *Propionibacterium* spp. (Kariluoto et al. 2014) are an example of other folate producing bacteria. *Aureobasidium pullulans*, *Candida lusitaneae*, *Kluyveromyces marxianus*, *Pseudozyma* spp., *Rhodotorula glutinis*, *Rhodotorula laryngis* and *Saccharomyces cerevisiae* are examples of producer yeasts (Korhola et al. 2014; Kariluoto et al. 2014). The only fungus described as folate producer is *A. gossypii* (Serrano-Amatriain et al. 2016) which was a target of metabolic engineering in order to discover an alternative to synthetic production of the vitamin.

The concentration of folate produced by the mentioned micro-organisms differs between strains and depends on the medium/matrix (Laiño et al. 2013), growth conditions (Albano, Silvetti, and Brasca 2020) and bacteria genetics. In the study performed by Albano and colleagues (2020), folate production by an *E. faecium* strain grown in M17 at 37ºC overnight reached 116642.9 ± 1.3 ng mL^{-1}. The highest concentrations were produced by *E. faecium*, *Lc. lactis*, *Lb. plantarum*, *L. delbrueckii* subsp. *bulgaricus* and *S. thermophilus*.

2.7. Biological Activity

Folate plays an important role as coenzyme in cellular metabolic pathways. This vitamin participates in the metabolism of nucleotides in DNA and RNA synthesis, cellular division, repair and methylation. Folate is also involved in the metabolism of amino acids, for example in the conversion of homocysteine into methionine, necessary for neurotransmitters and phospholipids synthesis (Acevedo-Rocha et al. 2019; Albano, Silvetti, and Brasca 2020; Hemery et al. 2020; Thompson et al. 2020). The demand for this vitamin is high during the early stages of pregnancy concerning DNA synthesis and methylation reactions due to the development of the embryo (Tamene et al. 2019).

In addition to this relevant role, the intake of this vitamin is also associated with some health benefits, which include a general improvement of folate status, decrease of homocysteine in serum and other indicators of inflammation, prevention of neural tube defects, megaloblastic anaemia, pancreas, breast and colorectal cancer, Alzheimer disease and coronary disease and helps in the treatment of some diseases like colitis, mucositis and colon cancer (Albano, Silvetti, and Brasca 2020; LeBlanc et al. 2020; Tamene et al. 2019). Beyond these benefits, the intake of foods fermented by folate producing bacteria demonstrated anti-inflammatory associated effects preventing the development of a wide range of inflammatory diseases, namely of inflammatory bowel diseases (IBD) due to the modulation of the intestinal microbiota and thus leading to the improvement of the intestinal barrier. Moreover, the reduction of oxidative stress and improvement of immune response were also described, helping in the treatment of patients with inflammatory diseases, like intestinal mucositis, and preventing secondary effects of their medication (Levit et al. 2019; LeBlanc et al. 2020).

2.8. Dietary Requirements in Humans

The Food and Nutrition Board, Institute of Medicine (Institute of Medicine and Food and Nutrition Board 1998) defined values for adequate intake (AI) and recommended dietary allowances (RDAs) of folate. These values differ from values established later (as shown in Tables 1 and 2), with 9 to >70 years-old having separate RDA values for men and women. Up to 1-year-old AI was 65 μg day^{-1} in the first semester and 80 μg day^{-1} in the second. RDAs from 1 to >70 years-old (independently of gender), and of pregnant and lactating women were equal to RNI values later set by FAO and WHO (2004), showed in Table 6.

FAO and WHO (2004) established values for estimated average requirement (EAR) and recommended nutritional intake (RNI). RNI values were generally higher than EAR. Up to 1-year-old RNI was 80 μg day^{-1}, from 1 to 10 years-old RNI values increase up to 400 μg day^{-1} remaining equal until the end of life. The highest RNI values, 600 or 500 μg day^{-1}, were set for pregnant women and lactating women, respectively (Table 6).

The EFSA Panel on Dietetic Products, Nutrition and Alergies (NDA) (2014) defined values for average requirement (AR) and population reference intake (PRI) of vitamin B9. Infants with less than one year-old do not have AR or PRI values but have an AI set at 80 μg day^{-1}. AR values start at 90 μg day^{-1} for 1-3 year-old children and rise up to 250 μg day^{-1} for those aged 15 or over; during lactation folate average requirement was set at 380 μg day^{-1}. PRI values increase from 120 to 330 μg day^{-1} between the ages of 1 and 15 and remains at the highest value throughout life. The highest PRI values, 600 or 500 μg day^{-1}, were set for pregnant women and lactating women, respectively (Table 7). AR and PRI values were slightly lower than the previous EAR and RNI values, respectively, set by FAO and WHO (2004). Exceptions are the PRI values for pregnant and lactating women, which are equal.

Table 6. Vitamin B9 recommended nutritional intake (RNI) (FAO and WHO 2004)

Group	RNI (μg day^{-1})
0-6 months	80
7-12 months	80
1-3 years	150
4-6 years	200
7-9 years	300
10-18 years	400
19-65 years	400
> 65 years	400
Pregnant women	600
Lactating women	500

Table 7. Vitamin B9 adequate intake (AI), average requirement (AR) and population reference intake (PRI) (EFSA Panel on Dietetic Products, Nutrition and Alergies (NDA) 2014). (1 μg DFE = 1 μg of food folate = 0.6 μg of acid folic from fortified food or as supplementation)

Group	AI (μg DFE day^{-1})	AR (μg DFE day^{-1})	PRI (μg DFE day^{-1})
7-11 months	80	-	-
1-3 years	-	90	120
4-6 years	-	110	140
7-10 years	-	160	200
11-14 years	-	210	270
≥15 years	-	250	330
Pregnant women	-	-	600
Lactating women	-	380	500

2.9. Cellular Absorption

Natural folates (polyglutamates) have lower absorption rates than folic acid (monoglutamates); polyglutamates need to be hydrolysed to monoglutamates to be absorbed. Hydrolysis is performed by glutamate carboxypeptidase II in the intestinal mucosa. The absorption may occur in the duodenum although it largely occurs in the jejunum, by a proton-coupled folate receptor (PCFT), which is a pH dependent mechanism. Absorption may also be mediated by other transporters, such as folate binding protein (FBP) and other folate receptors, proton-coupled folate transporter (identified in folate malabsorption patients), organic anion transporters and reduced folate carrier. During transportation there is a reduction and methylation of all folate forms to 5-methyltetrahydrofolate (5-MeTHF). This is the main form of the circulating vitamin that also enters the cells by transportation mediated by carriers or receptors.

Bioavailability is affected by some factors like the vitamin form, genetic variants of enzymes and the lifestyle and health status of the individuals. The bioavailability is also determined by its absorption, which mainly occurs in jejunum, consequently leading to a lower bioavailability of natural folates than folic acid (Rossi et al. 2016; Snauwaert et al. 2015; Sybesma, 2003; Zempleni et al. 2013).

2.10. Vitamin Deficiency

The lack of an optimal intake of folate and consequent deficiency are associated with a range of negative effects in human health. The most common health problems associated with folate deficiency are megaloblastic anaemia, caused by the decrease of cell division rate, inflammatory bowel disease and high levels of homocysteine in plasma. During pregnancy this deficiency may lead to adverse effects, due to the high levels of homocysteine, and neural tube defects (Hemery et al. 2020; LeBlanc et al. 2020). The lack of the vitamin also leads to hypomethylation of DNA, different types of cancer, neurologic disorders, depression and cardiovascular problems.

Folate hypovitaminosis is not unusual, turning this deficiency into a subject of global health concern. There are two approaches to address this problem, which are supplementation and food fortification. Supplementation is the intake of a high dose of the vitamin, in the chemically synthetized form (folic acid). Food fortification consists in the increase of vitamin B9 concentration through the addition of folic acid, or the production of functional foods fermented with folate producing microorganisms. Several countries have implemented food fortification programs to counteract B9 deficiency (Hemery et al. 2020; Serrano-Amatriain et al. 2016; Thompson et al. 2020) and some associated effects, such as the establishment of mandatory addition of folic acid to flour in United States of America, Canada and Argentina (Hemery et al. 2020; Khalili et al. 2019), and to food grains in USA, United Kingdom, New Zealand and Australia. The consumption of cereal based staple foods, fermented with microorganisms, is usual in developing countries, mostly by low income population in some African countries (Guyot 2012). Vitamin status improved due to the implementation of these measures and led to a decrease of related health problems. Despite of the observed benefits, some countries did not adopt the fortification with folic acid since, unlike natural folates, folic acid can be prejudicial. Folic acid may cause adverse effects such as masking precocious haematological manifestations of vitamin B12 deficiency - an excessive exposure of the foetus could favor methylenetetrahydrofolate polymorphism associated to a range of debilitating diseases, alteration of hepatic dihydrofolate reductase activity and promotion of cancer (Albano, Silvetti, and Brasca 2020).

This leads to a necessity of finding natural sources and functional foods bio-enriched with folate to prevent folate deficiency and its effects, without risking any other adverse effects.

2.11. Food Sources

2.11.1. Natural Sources

Green leaves, vegetables, nuts, fruits, liver, egg yolk, cereals, fermented milk products, meat and liver are natural sources of folate. The vitamin content is usually high in products fermented by LAB, such as fermented milk products (Albano, Silvetti, and Brasca 2020; Thompson et al. 2020).

Table 8. Examples of reference values of folate content in foods (U.S. Department of Agriculture 2016)

Sources	Peanuts	Cauliflower	Potatoes	Chicken		Milk	Yoghurt	Cheese
				Liver	Eggs			
Folate concentration (μg *per* 100g)	245	57	38	577	209	5	7-77	up to 63

2.11.2. Functional Foods

Fermented foods with increased folate concentration have already been developed using substrates such as barley, pseudo cereal, oats, milk, goat milk (Albano, Silvetti, and Brasca 2020; Khalili et al. 2020), potatoes, quinoa flour (Carrizo et al. 2020), teff (Tamene et al. 2019), vegetables (Kim et al. 2020) and a mixture of cauliflower and beans (Thompson et al. 2020).

Kariluoto and co-workers (2014) explored oat and barley matrices. These researchers isolated yeasts and bacteria from oat and rye products and folate production by these microorganisms was investigated and tested in cereal matrices. *S. thermophilus* and *L. reuteri* were the LAB capable of bio-enriching these matrices, with *S. thermophilus* presenting the best performance. LAB increased folate concentration more than 0.020 μg g^{-1}, achieving 0.030 μg g^{-1} following two weeks of storage. In spite of the scarce folate biosynthesis, the authors proved that oat and barley matrices may be interesting to develop functional products.

Carrizo and colleagues (2017) investigated the viability of amaranth based foods as suitable delivery vehicles of folate. The authors isolated 29 different LAB strains from spontaneous amaranth sourdough (SAS) and amaranth grain (AG). Among the isolates, *L. plantarum* CRL2106 demonstrated the highest folate production of 138 ± 7.5 μg L^{-1} and the second highest phytase activity, of 730 ± 25 U mL^{-1}. The highest phytase activity observed was 1041 ± 25 U mL^{-1}, for *E. durans* CRL 2122. With this work, it was proven that amaranth based foods may be fermented with folate producing LAB increasing their vitamin contents and shelf-life, by decreasing the pH values, and decreasing vitamin degradation by phytate. Thus, they are possible B9-biofortified foods.

Injera is a cereal-based fermented food manufactured with teff dough. Tamene and colleagues (2019) collected 60 samples of teff, following the first stage of fermentation that lasted 3 to 4 days. One hundred and sixty-two potential folate-producing LAB were isolated but only 19 of these were able to grow in a folate free culture medium, in 24 h at 30ºC. The biosynthesis of folate by these isolates ranged between 1 to 43 $\mu g\ L^{-1}$; the four highest producers were strains of *L. plantarum*, producing more than 30 $\mu g\ L^{-1}$. In order to evaluate the impact of the ingestion of *L. plantarum* P2R3FA, the highest B9 producer, *in vivo* studies were conducted, in which the lyophilised form of the bacterium was incorporated into animals' feed. The experiment comprised a total of thirty rats, fed with a folic acid deficient diet (FADD) for 30 days. Subsequently, 10 were sacrificed, 10 were fed with FADD fortified with 2 mg of synthetic folic acid *per* kilogram of diet, and 10 were fed with FADD supplemented with freeze dried *L. plantarum* cultures (containing a total content of 0.25 mg of folate *per* kilogram of diet), for 28 days. Despite no significant differences in the total intake and weight gained were observed between the two groups (fed with FADD supplemented with folic acid or folate producing *L. plantarum*), the intake of the vitamin was significantly different due to its concentration in the different diets. The concentration of the B9 vitamin in the erythrocytes and serum of the rats of the different groups was significantly different, as expected. Comparing the total intake of the vitamin with the concentrations of the vitamin in the mice erythrocytes and serum, it was observed a higher percentage of absorption in the group fed with the natural form of the B9 vitamin produced by *L. plantarum* cultures. In the mentioned study, it was demonstrated that the intake of both synthetic and natural forms of the vitamin improved the folate status. The intake of a diet mixed with folate producing LAB enhanced folate content more than 10 $nmol\ L^{-1}$ in serum, and more than 100 $nmol\ L^{-1}$ in erythrocytes, after depletion of the vitamin. Since the authors did not analyse the folate concentration in the teff dough, neither administered this putative B9 biofortified food to the rats, it was not possible to confirm the product functionality. Nevertheless, teff dough may be a probable functional food, fermented with folate producing LAB.

Pasta made with quinoa flour and fermented with a mix of LAB was developed by Carrizo and colleagues (2020). Quinoa flour was mixed with four different LAB combinations, separately, to prepare doughs, which were subsequently fermented for 24 h at 30ºC. Following fermentation, the concentration of folate and phytate were evaluated, leading the authors to select the mixture of *L. plantarum* CRL 2104 and *L. plantarum* CRL 1964 to proceed to the subsequent stages of the experiment. The combination of both strains produced the highest concentration of folate, 1.6 ± 0.2 µg g^{-1}, and the lowest concentration of phytate, 17.4 ± 1.3 mg g^{-1}, almost 50% of phytate concentration in the control (32.4 ± 0.5 mg g^{-1}), demonstrating phytase production). The selected dough was subsequently used to prepare a B9 bio-enriched quinoa pasta, to be used in an *in vivo* experiment to assess the effectiveness of developed food in preventing the folate deficiency. In this sense, the bio-enriched pasta was employed as a treatment in a bioassay to feed mice, in which the other treatments were: (i) a deficient diet; (ii) a control diet, equal to the deficient diet but enriched with synthetic vitamin; (iii) a pasta pH control diet; (iv) a negative control diet, fermented with a non-vitamin producer bacterium; (v) and a control pasta diet enriched with synthetic vitamin. The folate content of the developed biofortified pasta was significantly higher than the negative control diet and phytate content was lower than in deficient and control diets. Despite no differences being observed in the amount of food consumption, the mice fed with enriched and bio-enriched diets gained more weight than the other groups. The group of mice fed with the bio-enriched pasta presented higher hematological values, along with significantly higher levels of folate in blood and liver than the mice in all the other groups.

Nonetheless, concerning folate content determined in the mice kidney, the authors did not observe differences between the animals administered with the novel bio-fortified pasta and the mice fed with enriched pasta (v). There were histological benefits in the enriched groups, having more intestinal villi and with higher length, being the longest observed in the bio-enriched group. In conclusion, the authors were not only able to develop a functional food, formulated with quinoa flour, but also proved that their product presented health benefits concerning hematological, histological and folate status parameters.

Tuber-based foods with increased folate content synthesized by potential probiotics were developed by Mosso and co-workers (2018). The concentrations of total folate ranged from 29.4 µg L^{-1} (*L. fermentum*) up to 139.7 µg L^{-1} (*L. sakei*). *Solanum tuberosum* ssp. *andigena*, *Oxalis tuberosa* and *Ullucus tuberosus* were converted into purees, inoculated, individually, with five selected LAB, and incubated at 37ºC for 24 h. Those LAB demonstrating the best performance in this experiment were also inoculated in two potato mixtures, containing potato puree with amaranth flour and potato puree with amaranth and chia flour, in the same conditions as abovementioned. All the strains showed a significant growth in the tuber purees. The highest concentrations of folate were produced by *L. sakei* 2T3MS8 and *L. sakei* 2T2MM10, yielding a vitamin content between 0.760 and 1.484 µg g^{-1}. The fermentation of potato mixtures, possessing greater nutritional value than the tubers, led to higher concentrations of folate, in which the highest increase was obtained in the mixture containing potato puree, aramanth and chia, with a B9 vitamin improvement of 89% by *Lb. sakei* 2T3MS8 and 95% by *Lb. sakei* 2T2MM10, reaching 1.9 µg g^{-1}. In this study, a novel group of bio-enriched foods was developed, being demonstrated that the bacteria employed were potential probiotics. The two highest-folate producing isolates, *L. sakei* 2T3MS8 and *L. sakei* 2T2MM10, were deposited in the Culture Collection of the Centro de Referencia para Lactobacilos, in Argentina (CERELA), as *L. sakei* 2209 and *L. sakei* 2210.

Thompson and colleagues (2020) developed a mixture of cauliflower and beans with enhanced concentration of folates. The experimental set up included three matrices, namely, raw cauliflower, cooked white beans and a mixture of raw cauliflower and cooked white beans to which 2 g of sea salt were added. All the matrices were inoculated with 1% (w/w) of four different strains of *L. plantarum*, independently, and incubated for 44 hours at 30ºC. Analysis performed in the mixture of cauliflower and white beans showed an increase in the folate content ranging from 32 to 60%. The highest increase was obtained in the fermentation performed with *L. plantarum* 299v, isolated from the human gastrointestinal tract, reaching 58.8 ± 2.0 μg 100 g^{-1} of fresh weight. The production of this vitamin was considered extracellular since no cell lysis step was executed and inoculated strains maintained high stability when frozen. In the mentioned study, the authors demonstrated the formulation of a novel fermented food, containing an increased folate content along with an extended shelf-life, due to the low pH value, being a potential modulator of the intestinal microbiota, as a direct source of the vitamin on a regular basis.

Kim and colleagues (2020) analysed the whole genome and metabolic features of 34 *L. sakei* (only four strains isolated from kimchi) during fermentation of kimchi. The complete gene set of folate biosynthesis was found in the genomes of all the isolates. Transcriptomic analysis of the strains showed that folate metabolic biosynthetic pathway was significantly expressed during fermentation, demonstrating that kimchi may be considered a B9 biofortified food.

Laiño and co-workers (2012) developed a bio-enriched fermented milk, with LAB isolated from artisanal yogurts. In the mentioned study, 92 LAB, *L. delbrueckii* subsp. *bulgaricus* (41) and *S. thermophilus* (51), were tested for folate production, followed by screening for the vitamin synthesis in a folate-free medium. The three highest producers, *L. delbrueckii* subsp. *bulgaricus* CRL 863, *S. thermophilus* CRL 803 and *S. thermophilus* CRL 415, were chosen to inoculate milk (reconstituted from non-fat powdered milk), which was fermented for 24 h at 37° and 42°C. All the treatments showed improved folate concentrations in comparison with control at 24 h. The highest producer, at both temperatures, was *S. thermophilus* CRL 803, increasing folate content in 39.5% at 37°C and 175% at 42°C in 24 h, achieving *ca.* 80 μg L^{-1} (in both conditions).

Pacheco Da Silva and colleagues (2016) developed fermented goat milks with LAB isolated from goat milk and commercial cheese samples. A total of 21 isolates were classified as folate producers, yielding from 23 to 498 μg L^{-1} in a folate-free medium, out of which seven were selected to inoculate goat milk. *S. thermophilus* FP268 was the first to reach high concentrations of folate, taking 6 h to synthesize almost five times the initial concentration. However, the highest producer was *Lc. lactis* subsp. *lactis* FP368, with a maximum production of 313 ± 81 μg L^{-1} of folate, after 24 h of fermentation. Nevertheless, *S. thermophilus* FP268 was an interesting isolate due to its capacity of rapid production of high folate concentration.

Bio-enriched fermented milk (BeFM) was a functional food formulated by Jiao and colleagues (2020). Sterile milk was inoculated with two *Lc. lactis*, independently, of which one was a high folate producer and the other strain a modest producer and the fermentative process was conducted for 24 h at 37ºC. When compared with the unfermented milk only the BeFM with *Lc. lactis* KLDS4.0325, the highest folate producer, presented a folate concentration of 30.59 μg *per* 100 mL. The B9 vitamin content in BeFM was adjusted to 0.6 μg mL^{-1}, which was further evaluated in an experiment with mice model. These *in vivo* bioassays comprised seven treatments, in which one group was fed with a control diet during the whole experience, while the other six were fed with a folate deficient diet (FADD) for the first 14 days and for the following 28 days with FADD, FADD with 0.6 μg of folic acid day^{-1}, FADD with 1.2 μg of folic acid day^{-1}, FADD with fermented milk, FADD with BeFM, and FADD with 0.6 μg of folic acid day^{-1} enriched fermented milk. At the end of the 42 days, blood and samples of colonic content of the mice were collected to be analysed. Plasma analysis showed that, except for milk fermented by a scarce folate producing *L. lactis* and FADD with 0.6 μg of folic acid day^{-1}, all the treatments were able to reverse the increase of homocysteine caused by folate deficiency. Haematological analysis demonstrated that in mice fed with BeFM, folic acid concentration in the whole blood was significantly lower when compared with mice fed with control diet, despite no significant differences being observed in folic acid concentration in liver, neither in 5-MeTHF (5-methyltetrahydrofolate) content in blood and liver. The analysis of colonic content showed that the consumption of BeFM contributed to an increase of Bacteroidetes and a decrease of Actinobacteria, reversing folate deficiency effect. The total concentration of the vitamin in mice fed with BeFM was significantly lower than in mice fed with control diet, but significantly higher than in mice fed with deficient diet, fermented milk, and 0.6 μg of folic acid day^{-1}. The developed functional product improved folate status in mice with folate deficiency, which did not occur with milk fermented by a scarce folate producing *Lc. lactis*.

Laiño and his team (2013) developed an yogurt with enhanced folate concentration using previously characterized LAB (Laiño, Leblanc, and Savoy de Giori 2012). Five combinations of LAB were used to produce yogurt (42ºC for 6 h), due to their high folate production in milk (being the highest yield 110.6 μg L^{-1}). The combination of LAB composed by *L. delbrueckii* subsp. *bulgaricus* CRL 871, *S. thermophilus* CRL 803 and *S. thermophilus* CRL 415 was the one producing the highest concentrations of folate in yogurt, nearly 180 μg L^{-1}.

A folate bio-fortified yoghurt, fermented with LAB, was developed by Khalili and colleagues (2020). Heat-treated cow milk was inoculated with five different mixtures of LAB, comprising commercial cultures and LAB isolated from dairy sources, fermented for 6 h at 42ºC, and then stored for 21 days at 4ºC. All the mixtures contained *S. termophilus* and *L. delbrueckii* subsp. *bulgaricus* as the starter cultures (also present in conventional yoghurt). Four of them also contained one or two of the pure strains of probiotic bacteria, *L. acidophilus* LA-5 and *Bifidobacterium animalis* subsp. *lactis* BB-12 (1), *L. plantarum* 15HN (2), *Lc. lactis* subsp. *lactis* 44Lac (3) and *L. plantarum* LAT BY PL (4). The enumeration of bacteria performed throughout storage showed that the probiotic *L. plantarum* 15HN inoculated into yoghurt remained viable for 21 days (9.09 ± 0.61 CFU mL^{-1} at day 1, and 9.33 ± 0.60 CFU mL^{-1} after 21 days). The concentrations of two derivatives of folate were analysed in milk and throughout storage. The concentration of 5-methyltetrahydrofolate (5-MeTHF) at the end of storage was higher in the yoghurt comprising the probiotic *L. plantarum* LAT BY PL, reaching 621.33 ± 18.58 μg L^{-1}. The highest concentration of 5-formyltetrahydrofolate (5-FmTHF) at the end of storage was 747.33 ± 2.30 μg L^{-1}, in the yoghurt with the probiotic *Lc. lactis* subsp. *lactis*. However, the highest concentrations of these compounds were detected in the yoghurt containing *L. plantarum* 15HN, reaching 642 ± 46.50 μg (5-MeTHF) L^{-1} in the 14th day, and 1166.70 ± 57.73 μg (5-FmFTH) L^{-1} in the 7th day. Khalili and colleagues were able to develop a functional food with high fortification provided by *L. plantarum* 15HN.

Albano and colleagues (2020) developed a bio-enriched cheese, starting by the identification of folate producers through the vitamin quantification assay, among 70 LAB strains belonging to a culture collection comprising *L. casei*, *L. delbrueckii* subsp. *bulgaricus*, *L. paracasei* subsp. *paracasei*, *L. plantarum*, *L. rhamnosus*, *Lc. lactis* subsp. *lactis*, *E. faecium*, *E. lactis* and *S. thermophilus*. The 47 identified producer strains were then analysed for determination of total, intra- and extracellular folate. The highest concentrations of total B9 vitamin detected for the five highest producing strains ranged from 89.50 ± 2.00 to 123,625.74 ± 8.00 μg L^{-1}. Goat milk was inoculated with each of these five strains (independently), namely *L. casei* VC199, *L. paracasei* subsp. *paracasei* SE160, *L. plantarum* VS513, *E. faecium* VC223, and *E. lactis* BT161, along with a thermophilic lactic culture. No significant differences of folate concentration were detected in control samples after 30 and 60 days of ripening. The cheeses obtained following 30 days of ripening presented significantly higher concentrations of folate, when compared to the control cheese, ranging from 40.31 to 86.68 μg $100g^{-1}$. The highest producer was *L. casei* VC199, increasing folate content in 30.29 μg $100g^{-1}$, corresponding to 54% of control content. After 60 days of ripening all the treatments showed significant differences, when compared to the control. The highest producer strain was *L. paracasei* subsp. *paracasei* SE160, reaching 122.58 μg $100g^{-1}$ of folate at the 60th day. This concentration represented an increase of folate level of 63.52 μg $100g^{-1}$, corresponding to 108% of control content. The three LAB inoculated in the cheeses which presented the highest folate concentrations at 30th and 60th days were *L. casei* VC199, *L. paracasei* subsp. *paracasei* SE160 and *L. plantarum* VS513.

The abovementioned bio-enriched fermented foods and their vitamin B9 content are presented in Table 9.

Table 9. Folate content in LAB fermented food products

Biofortified Foods	Microorganism	Vitamin B9 content	Reference
Amaranth based foods	*L. plantarum* CRL2106	138 ± 7.5 µg L^{-1}	Carrizo et al. 2017
Oat and Barley matrices	*S. thermophilus*, *L. reuteri*	0.030 µg g^{-1}	Kariluoto et al. 2014
Tef dough	*L. plantarum* P2R3FA	0.25 µg g^{-1}	Tamene et al. 2019
Quinoa flour	*L. plantarum* CRL 2104 and *L. plantarum* CRL 1964	1.6 ± 0.2 µg g^{-1} of dough	Carrizo et al. 2020
Quinoa pasta		1.3 ± 0.1 µg g^{-1} of pasta	
Tuber-based foods (Purees)	*L. sakei* 2T3MS8 and *L. sakei* 2T2MM10	0.760 to 1.484 µg g^{-1}	Mosso et al. 2018
Potato based foods (Potato puree with amaranth and chia)	*L. sakei* 3T3M3	1.9 µg g^{-1}	
Cauliflower and beans	*L. plantarum* 299v	58.8 ± 2.0 µg 100 g^{-1}	Thompson et al. 2020
Kimchi	*L. sakei*	-	Kim et al. 2020
Fermented milk	*S. thermophilus* CRL 803	~80 µg L^{-}	Laiño, Leblanc, and Savoy de Giori 2012
Goat fermented milk	*Lc. lactis* FP368	313 ± 81 µg L^{-1}	Pacheco Da Silva et al. 2016
Fermented milk	*Lc. lactis* KLDS4.0325	30.59 µg *per* 100 mL	Jiao et al. 2020
Yogurt	*L. delbrueckii* subsp. *bulgaricus* CRL 871, *S. thermophilus* CRL 803 and *S. thermophilus* CRL 415	~180 µg L^{-1}	Laiño et al. 2013
Yoghurt (cow milk)	*L. plantarum* 15HN	642 µg (5-MeTHF) L^{-1} 1166.70 µg (5-FmFTH) L^{-1}	Khalili et al. 2020
Cheese	*L. casei* VC199; *L. paracasei* subsp. *paracasei* SE160	86.68 µg 100 g^{-1} (30th day); 122.58 µg 100g^{-1} (60th day)	Albano, Silvetti, and Brasca 2020

2.12. Stability in Foods

Vitamin B9 stability, product matrix and storage conditions are crucial factors after fortification or bio-enrichment of the foods, to determine the success of the process, since the mentioned vitamin degrades when exposed to oxygen, increased temperatures and humidity.

Product matrix may have an effect on folate content depending on its composition. The presence of phytate leads to a decrease in the concentration of the vitamin. In some studies of bio-enriched products it was demonstrated that microbial phytase activity (activity that degrades phytate), is an asset to the vitamin content.

Storage conditions such as temperature, humidity and packaging, were assessed by Hemery and colleagues (2020). Vitamin stability in folic acid fortified flour was tested during storage at 25 and 40ºC, 65 and 85% of relative humidity and in paper and aluminium/PET bags. Vitamin stability was higher in the aluminium/PET bags, independently of the other conditions. In the natural folate bio-enriched flour, vitamin stability was also higher in the aluminium/PET bags, although, in the paper bags with high humidity there was an increase in folate content, which increased with the increase of temperature.

Permeability to oxygen of the package, temperature and humidity are important factors to vitamin stability during storage. Moreover, if the product is fortified, the type of packaging and storage conditions must be evaluated since in the appropriate conditions the vitamin content may increase. Hemery and colleagues (2020) observed an increase of natural folate content in flour stored in paper bags at 25 and 40ºC and 85% of relative humidity, after 60 days.

In the case of biofortified products, the type of folate synthesized, intracellular or extracellular, will have an important role according to the type of heat-processing. Heat-processing allows releasing intracellular folate by cell lysis, while in raw food intracellular vitamin content is not available for absorption, except if the bacterial cells lysis occurs in the digestive system. In this sense, the consumption of raw bio-enriched food with extracellular folate producing microorganisms is preferred since these microorganisms excrete the vitamin in the gastrointestinal tract, or colonise it, becoming a regular source of the vitamin.

2.13. Conclusion

Vitamin B9, biosynthesized as natural folate and derivatives, and chemically synthesised as folic acid, is essential. Chemical synthesis allows production with much higher yields and it is more economic. However, an excessive intake of this form of the vitamin is associated with negative health effects, contrarily to natural forms.

Functional foods bio-enriched by folate producing LAB are a safe strategy to counteract vitamin B9 deficiency that should be further explored.

3. Vitamin B12

3.1. Vitamin B12

The term vitamin B12 refers to all cobalt corrinoid compounds, including those from cobalamin group. Vitamin B12 was discovered in early 1920, by two American physicians, Minot and Murphy, as a cure for a liver disorder named pernicious anaemia, and they referred it as an "extrinsic factor" (Minot and Murphy 1926). After 20 years of research the "extrinsic factor" was finally isolated from the liver as a red crystalline compound by two pharmaceutical companies, one directed by Folkers at Merck in the USA, and the other by Smith at Glaxo in the UK, and it was designated vitamin B12 (Rickes et al. 1948). Vitamin B12 was also identified in milk powder, beef extract, and in several bacteria genera culture broths (Rickes et al. 1948).

After these discoveries, the biologically active coenzyme forms of pseudo-vitamin B12 and vitamin B12 were crystallized (Barker, Weissbach, and Smyth 1958). Then, the three-dimensional structure of vitamin B12 (cyanocobalamin) was determined and five years later, the structure of coenzyme B12 (adenosylcobalamin) (Hodgkin et al. 1955, 1956, 1957; Lenhert and Hodgkin 1961). A second biologically active form of vitamin B12, methylcobalamin (MeCbl) was also determined and this form is a cofactor of methionine synthase (Guest et al. 1962). Following these findings, more cobalamin-dependent enzymes and their action mechanisms have been identified and studied (Martens et al. 2002).

3.2. Chemical Structure and Properties

The term vitamin B12 represents, as previously mentioned, a cobalt corrinoid family, particularly the cobalamin group, and is widely found in two naturally occurring forms that are referred to as coenzyme B12: adenosylcobalamin (AdoCbl) (Figure 6) and methylcobalamin (MeCbl). Coenzyme B12 has a molecular mass of 1,580 Da. Commonly, vitamin B12 also refers to cyanocobalamin (CNCbl), an industrially produced form, and the cyano group (CN) is obtained through an extraction procedure that removes this compound from bacterial cultures (Martens et al. 2002).

This molecule is constituted by a central element, a corrin ring containing four pyrrole ligands. The Cobalt (III) ion is chelated by nitrogen atoms donated by the four pyrrole rings. Two additional ligands are covalently linked to the cobalt ion, a lower (alpha) ligand, which is a N7-atom donated from 1,3-dimethyl-2-phenyl-2,3-dihydro-1H-benzoimidazole (DMBI), and an upper (beta) ligand, that can be a CN, adenosyl, methyl or hydroxy group, forming different analogues. All the acetamide groups are projected in the upper ligand direction, while all the propionamide groups are projected in the benziminazole group direction (Hodgkin et al. 1956; Martens et al. 2002).

Vitamin B12 appears in the form of dark red crystals or as a red crystalline powder, being odourless and tasteless. Its melting point exceeds 300ºC and when exposed to air it can absorb up to 12% of water, being hygroscopic (Laird and Molly 2014). In aqueous solution it is stable in a range of pH values between 4.0 to 6.5 (Taub and Lieberman 1953). Concerning stability, cyanocobalamin is the most stable form of the vitamin B12 analogues. It can be submitted to 120ºC during autoclave sterilization, is relatively stable at 100ºC for a few hours and is also stable in both air and dry forms. Methylcobalamin and adenosylcobalamin are stable in aqueous solution in dark conditions and can be submitted to 100ºC for 20 min (Jägerstad and Arkbåge 2003). In acidic or alkaline conditions, hydrolysis of the propionamide and acetamide side chains of vitamin B12 to carboxylic acid groups can occur, originating inactive products (Schnellbaecher et al. 2019). In highly acidic environment, hydrolysis of the phosphate bonds originates cobamamide and DMBI, which can be further degraded into other products (Bonnett 1963). At alkaline pH values and elevated temperatures, cyclization of the acetamide side chain of the B ring generates a γ-lactam, the biologically inactive dehydrovitamin B12 (Bonnett et al. 1957).

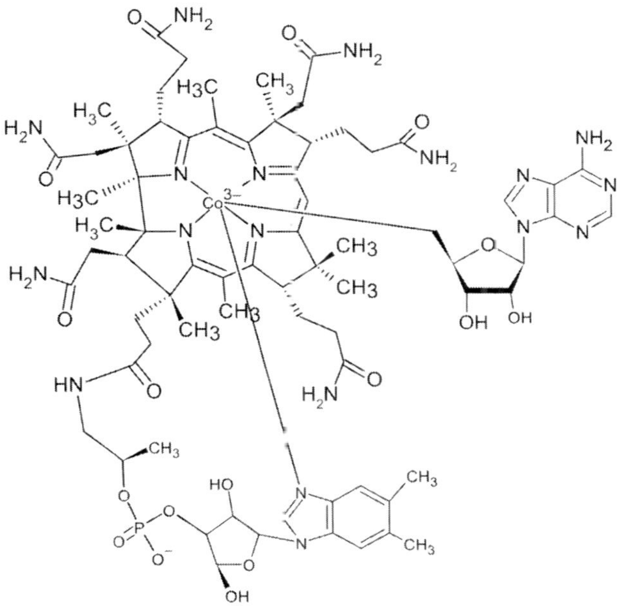

Figure 6. Chemical structure of coenzyme B12 (adenosylcobalamin) chemical structure (designed using ACD/ChemSketch; ACD/Labs 2019.2.1).

All forms of cobalamin are sensitive to ultraviolet and visible light (Demerre and Wilson 1956). In aqueous solution, cyanocobalamin is photolyzed to the most photostable form, hydroxocobalamin, and in seconds of UV-A exposure, the biologically active forms of cobalamin, methylcobalamin and adenosylcobalamin, are converted to hydroxocobalamin as well (Juzeniene and Nizauskaite 2013; Ahmad, Hussain, and Fareedi 1992). Thiamine was described as a vitamin B12 destroying agent. At high temperature and pH values, thiamine degradation products induce cobalamin degradation (Feller and Macek 1955). Moreover, the highest level of degradation occurs when thiamine and niacinamide are present simultaneously, since niacinamide induces the release of a substance from thiamine that interferes with cobalamin stability (Blitz, Eigen, and Gunsberg 1956). Finally, heavy metals, strong oxidizing and reducing agents can also affect the natural forms of cobalamins, including hydroxo-, aqua-, methyl-, or adenosylcobalamin (Jägerstad and Arkbåge 2003).

3.3. Biosynthetic Pathway

3.3.1. Vitamin B12 De Novo and Salvage Pathways

The biosynthesis of vitamin B12 can occur through the *de novo* or salvage pathways. Within *de novo* pathway, vitamin B12 can be synthetized via an oxygen-dependent or an oxygen-independent pathway. The aerobic mechanism may be found in microorganisms such as *Pseudomonas denitrificans* while the anaerobic mechanism may be found in *Bacillus megaterium*, *Propionibacterium shermanii* and *Salmonella* Typhimurium (Martens et al. 2002). At least 30 genes are involved in this vitamin *de novo* synthesis (Roth et al. 1993).

The vitamin synthesis begins in the tetrapyrrole synthesis pathway, with 5-aminolevulinic acid (ALA), synthesized either via C4 or C5 pathway, as a first committed precursor. Two molecules of ALA are then condensed to form porphobilinogen and this reaction is followed by the polymerization and cyclization of four porphobilinogen molecules, forming uroporphyrinogen (Fang, Kang, and Zhang 2017; Martens et al. 2002).

In both aerobic and anaerobic mechanisms, uroporphyrinogen is converted into precorrin-2 through a C-2 and C-7 methylation (Martens et al. 2002). This reaction is catalysed by a methyltransferase encoded by *cob*A gene (Warren et al. 2002). From this point on the pathways diverge since the cobalt ion is added at different phases (Raux et al. 1999; Martens et al. 2002). In the oxygen- independent pathway, cobalt ion is inserted into precorrin-2 (Martens et al. 2002). Chelatase CysG or independent precorrin-2 cobalt chelatase CbiK can catalyse this reaction in *Salmonella enterica,* and cobaltochelatase CbiX in *B. megaterium* (Warren et al. 2002). This early insertion results in a high number of cobalt-complexes as intermediates (Martens et al. 2002). The chelation in the oxygen-dependent pathway occurs nine steps later, and hydrogenobyrinic acid a,c-diamide is the substrate of a two component system cobalt chelatase that includes a fist component, CobN protein, and a second, CobS and T complex (Warren et al. 2002; Martens et al. 2002).

The ring contraction and subsequent C-20 removal in both pathways also presents differences. In the oxygen-dependent pathway, C-20 atom of precorrin 3-A is oxidized by molecular oxygen with release of C-20 as acetate, in a reaction catalysed by CobG protein, whereas in the oxygen-independent pathway the complexed cobalt ion may assume different valence states and allows oxidation and the C-20 atom is released as acetaldehyde. The two pathways converge when adenosylcobyric acid is converted into adenosylcobinamide (Martens et al. 2002). The last step involves the formation of an α-ribazole through the transference of a phosphoribosyl residue of nicotinic acid mononucleotide to DMBI, which is posteriorly linked to GDP-adenosylcobinamide. Guanosine 5'-monophosphate (GMP) is released in this reaction, giving origin to a coenzyme B12 molecule (Zayas and Escalante-Semerena 2007; Martens et al. 2002).

The salvage pathway is a system that allows bacteria and archaea to salvage exogenous cobinamide, being a more cost-effective way to produce cobalamin. *Escherichia coli* is an example of a microorganism that has lost *de novo* cobalamin production capacity across evolution but can synthetize it through the salvage pathway (Fang, Kang, and Zhang 2017). Gram-negative bacteria have a highly specific BtuBFCD system which translocates exogenous corrinoids from the environment, in an effective way. BtuB protein is a Ca-dependent transporter located in the outer membrane and its function is to deliver the corrinoid to BtuF protein, a periplasmic corrinoid-binding. BtuF delivers cobinamide to the ABC transporter, BtuCD, located in the inner membrane. Once inside the cell, cobinamide is adenosylated and converted into adenosylcobinamide. In bacteria, adenosylcobinamide serves as substrate for a bifunctional enzyme detaining kinase and guanylyltransferase activities (CobU, in *S.* Typhimurium or CobP, in *P. denitrificans*), and adenosylcobinamide-GDP is generated. The last two steps necessary for the production of adenosylcobalamin are identical to the *de novo* pathway (Escalante-Semerena 2007).

The vitamin B12 *de novo* and salvage pathways are represented in Figure 7.

3.3.2. Biosynthesis Regulation

Cobalamin biosynthesis is mainly regulated by riboswitches. Riboswitches are RNA elements situated in 5′-UTR of specific prokaryotic mRNAs that modulate the expression of certain genes in response to changing concentrations of metabolites. These regulatory elements are constituted by two functional structural domains, in which one serves as a natural aptamer, binding the target metabolite with high selectivity, and the other domain is an expression platform that suffers allosteric changes in RNA structure to control expression of the adjacent gene or operon (Nahvi, Barrick, and Breaker 2004). The mechanism of inhibition is activated when coenzyme-B12 concentrations are high. In that case, the nascent mRNA binds to a coenzyme-B12 molecule which induces an allosteric change in structure. This change allows the intrinsic terminator stem to form, the coding region of the mRNA is not generated and transcription stops at that point (Mandal and Breaker 2004).

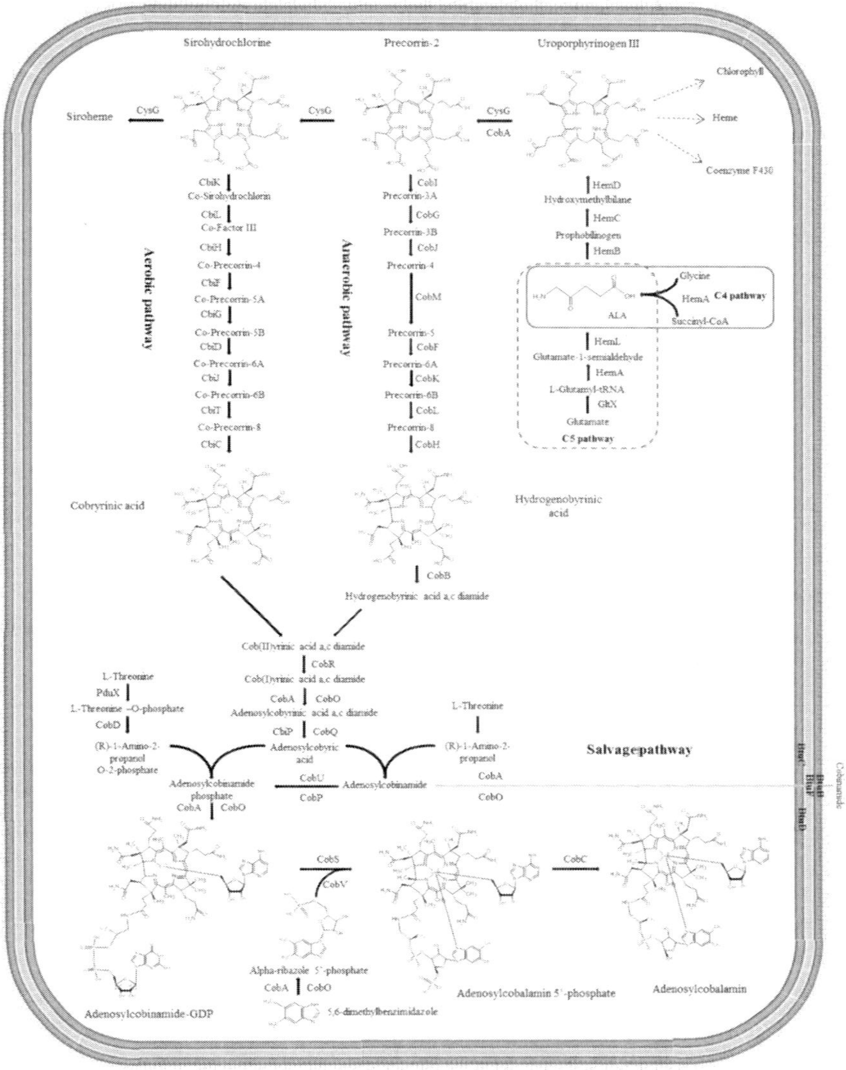

Figure 7. Vitamin B12 *de novo* and salvage pathways (designed using ACD/ChemSketch; ACD/Labs 2019.2.1; Adapted from Fang et al. (2017).

Riboswitches may be interesting targets for enhancing vitamin B12 production. *Propionibacterium* strain UF1 abundantly produces vitamin B12, and the biosynthesis is regulated by VB12, being regulated within the *cobA* operon through the cbiMCbl riboswitch. It was demonstrated that the ablation of the cbiMCbl riboswitch increased the production of vitamin B12 (Li et al. 2020).

3.4. Vitamin B12 Production

It was necessary more than 10 years and 100 researchers to achieve vitamin B12 full chemical synthesis, by Woodward and Eschenmoser in 1973. Vitamin B12 is commercially produced through fermentation processes using selected genetically modified microorganisms. The chemical synthesis of this compound is not a viable process since it evolves 70 steps, being highly complex and expensive (Martens et al. 2002).

To obtain high vitamin B12 production yields a random mutagenesis strategy is used being obtained by submitting microorganisms to mutagenic agents that comprise UV light, ethyleneimine, nitroso-methyluretane or N-methyl-N′-nitro-N-nitrosoguanidine and finally, selecting the most suitable strains. *Propionibacterium freudenreichii*, *Rhodopseudomonas protamicus*, *P. shermanii*, *Ps. denitrificans*, *Nocardia rugosa*, and *Rhizobium cobalaminogenum* are well characterized vitamin B12 producers (Kang et al. 2012; Bykhovskiĭ, Zaĭtseva, and Eliseev 1998).

Since *P. shermanii* and *Ps. denitrificans* strains produce vitamin B12 in high levels, these bacteria are frequently used in industrial production. *Propionibacterium* strains have the advantage of possessing a generally recognized as safe (GRAS) status. Nonetheless, *Ps. denitrificans* is more used in industrial settings (Martens et al. 2002).

The *Propionibacterium* strains employed in vitamin B12 production are microaerophilic and high yields are obtained when low oxygen concentrations are used. The fermentation is divided in two phases: in the first 3 days bacteria grow anaerobically to produce the vitamin B12 precursor cobamide (vitamin B12 without DMBI) and in the next 1 to 3 days aeration is initiated to allow the production of DMBI which will be linked to cobamide. During this process it is necessary to neutralize propionic acid to stabilize the pH value at 7. In the process using *Ps. denitrificans*, which has an oxygen-dependent growth, the culture is aerated throughout all the fermentation process.

3.5. Metabolic Engineering

Large scale industrial production of cobalamin occurs through fermentation. The most used strains have some setbacks, for instance, they often require long fermentation cycles and complex and expensive culture media. Most research has focused in optimizing fermentation process and employed mainly traditional methods, such as random mutagenesis. Therefore, there is the need to find suitable genetics systems for strain engineering and to extend the knowledge in metabolic engineering.

Some techniques can be used to enhance cobalamin production, including gene over-expression, to enhance a desired chemical reaction, gene downregulation or inactivation, to improve precursor supply or remove by-products and competing reactions, and protein engineering, to improve enzymes specificity, solubility and stability, limiting inhibition by a determined compound or product (Fang, Kang, and Zhang 2017). These techniques can be applied to strains with commercial value, and some studies to enhance cobalamin production through gene over-expression and introduction of heterologous genes have already been performed in *P. freudenreichii* (Piao, Kiatpapan, et al. 2004; Piao, Yamashita, et al. 2004).

The construction of heterologous biosynthetic pathways using model organisms can also be a promising strategy to produce cobalamin. *E. coli* has been widely employed as a microbial cell factory and it may be used as host for vitamin B12 heterologous production (Fang, Kang, and Zhang 2017). The *de novo* cobalamin synthesis in *E. coli* has already been reported in a few studies (Fang et al. 2018; Raux et al. 1996; Ko et al. 2014).

3.6. Vitamin B12 Biosynthesis by Lactic Acid Bacteria

Lactic acid bacteria comprise a heterogeneous group of microorganisms that have been used throughout the years as starter cultures for the production of fermented foods. Their importance relies on the capacity of improving safety, shelf-life, nutritional value, flavour, and overall quality of fermented products (LeBlanc et al. 2015).

The first lactic acid bacteria found to produce vitamin B12 was *Lactobacillus reuteri* CRL 1098, since this strain was described as capable of metabolizing glycerol in a B12 free medium (Taranto et al. 2003). The compound produced by this strain was identified as pseudocobalamin, a vitamin B12 analogue (Santos et al. 2007). A vast set of vitamin B12 synthesis genes were detected in *L. reuteri* ATCC 55730 (Hüfner et al. 2008). Furthermore, other strains including *L. reuteri* DCM 20016 (Santos et al. 2008), *L. reuteri* JCM1112, *L. reuteri* CRL 1324 and 1327 (strains isolated from human vagina) were identified as corrinoids producers (LeBlanc et al. 2011).

The genomic sequences of two other *L. reuteri* strains were revealed. *L. reuteri* F275 (type strain DSM 20016), isolated from human faeces, contained the cobalamin biosynthesis cluster associated with glycerol anaerobic catabolism, while *L. reuteri* 100-23, isolated from the mouse intestine, did not contain the *cob* or the glycerol metabolism genes. Of the LAB genomes studied thus far, only *L. reuteri* DSM20016/JCM1112 and the Biogaia strain *L. reuteri* ATCC55730 contained *pdu-cob* genes. Additionally, preliminary studies indicated that the cob-dpi cluster is present in *Lactobacillus coryniformis* and *Lactobacillus murinus* (LeBlanc et al. 2011). Some cobalamin biosynthesis genes were also shown to be present in *Lactobacillus buchneri* ATCC 11577, *Lactobacillus hilgardii* ATCC 8290, *L. coryniformis* KCTC 3167, *L. coryniformis* KCTC 3535 and *Lactobacillus brevis* ATCC 27305 strains (Capozzi et al. 2012).

It has already been shown that *L. coryniformis* strains CECT 571 and 394, isolated from goat milk, could produce reuterin through an enzymatic pathway that requires the presence of the cobalamin-dependent glycerol dehydratase enzyme. Moreover, *L. coryniformis* strain CECT 571 could also produce cobalamin (R Martín et al. 2005). Considering these findings, a study was performed using *L. coryniformis* CRL 1001 strain, in which it was observed that the testing strain could produce corrinoids with cobalamin activity, mainly pseudocobalamin. Furthermore, it was possible to conclude, through genomic studies, that the complete anaerobic biosynthesis pathway of coenzyme B12 was present. This was the first facultative heterofermentative strain described with biosynthetic capacity (Torres et al. 2016).

In a previous study, in which LAB isolated from Japanese pickles were investigated, *L. coryniformis* and *L. plantarum* strains revealed a vitamin B12 production of 1.8 and 2.0 mg L^{-1}, correspondingly, in a vitamin B12-free medium. In addition, one strain of *P. pentosaceus* and another of *L. brevis* also showed production capacity, even though at lower levels (0.4–0.6 mg L^{-1}) (Masuda et al. 2012). *L. plantarum* strains had already demonstrated vitamin B12 production potential, since it was observed that strains isolated from the fermented food kanjika could synthetize this vitamin (Madhu et al. 2010). In a phenotypic and genotypic screening of human-originated lactobacilli concerning vitamin B12 production capacity, two *L. plantarum* strains (BHM10, isolated from human milk, and BCF20, isolated from child faeces) were identified as producers after the *cbiK* gene was detected (Bhushan, Tomar, and Mandal 2016).

The research for more cobalamin producers has contributed to the expansion of this group to new species. A strain isolated from healthy breast-fed infants fresh faecal samples, identified as *L. fermentum* CFR 2195, revealed vitamin B12 production capacity, achieving a yield of 29.45 $ng_{vitamin\ B12}\ g^{-1}$ of the dry biomass (Basavanna and Prapulla 2013). Furthermore, a gene cluster involved in cobalamin biosynthesis was identified in *Lactobacillus rossiae* DSM 15814, with high degree of homology with the *L. reuteri* 20026 cluster. This suggests that this strain has vitamin B12 *de novo* synthesis capacity (De Angelis et al. 2014).

Enterococcus spp. with cobalamin production capacity were recently reported. Eight vitamin B12 producing strains were isolated from infant faeces, with five belonging to *Enterococcus* genus. Three isolates were identified as *E. faecium* and two as *Enterococcus faecalis*. Of the remaining isolates, two corresponded to *L. casei* and one to *L. plantarum*. The results revealed that *E. faecalis* LZ83 and *E. faecium* LZ86 strains presented the highest cobalamin production and phylogenetic analysis revealed high similarity with *E. faecalis* ATCC 19433 and *E. faecium* ATCC 19434 commercial strains, correspondingly. Both strains from the study have shown to produce active forms of cobalamin (Li et al. 2017).

3.7. Biological Activity in Humans and Animals

Vitamin B12 presents a biological role in humans and animals since it is required by two enzymes, namely, (R)-methylmalonyl-CoA mutase and methionine synthase (Martens et al. 2002).

The enzyme (R)-methylmalonyl-CoA mutase intervenes in propionyl-CoA metabolism that is obtained through degradation of thymine, valine, methionine and odd-chain fatty acids. In this process, propionyl-CoA suffers a carboxylation to (S)-methylmalonyl-CoA form and is epimerized to the (R)-methylmalonyl-CoA. A rearrangement performed by adenosylcobalamin-dependent (R)-methylmalonyl-CoA mutase occurs and succinyl-CoA is formed, being integrated in the tricarboxylic acid cycle. When vitamin B12 is absent, methylmalonylCoA intermediate accumulates and a hydrolase cleaves the D-isomer to coenzyme A and methylmalonic acid, originating acidosis. On the other hand, propionyl-CoA levels increase and it condenses with oxaloacetic acid to form 2-methylcitric acid, an aconitase inhibitor (Martens et al. 2002).

Methionine synthase, a MeCbl-dependent enzyme, catalyzes homocysteine methylation through the transference of a methyl group from 5-methyltetrahydrofolate, originating methionine and tetrahydrofolate (Banerjee and Matthews 1990). In cases of vitamin B12 deficiency, methionine levels remain normal, which may be explained by an increased activity of an enzyme betaine-homocysteine methyltransferase, that neither requires cobalamin nor folate (Allen, Stabler, and Lindenbaum 1993). Although the level of methionine is unchangeable, the deficiency of this vitamin induces the increase of methyltetrahydrofolate and the decrease of tetrahydrofolate levels. This results in low levels of folates that are used in DNA synthesis and it is called methylfolate trap. These events have been related with diseases like megaloblastic anaemia (Martens et al. 2002).

3.8. Dietary Requirements in Humans

The Recommended Dietary Allowance (RDA) for vitamin B12 is defined as the necessary amount of vitamin to satisfy the necessities of healthy individuals in a specific stage of life. The RDA for adults and other age groups was established by the Food and Nutrition Board of the Institute of Medicine (IOM) (Institute of Medicine and Food and Nutrition Board 1998). The RDA for different age groups and specific conditions is presented in Table 10.

Table 10. Recommended dietary allowance (RDA) for vitamin B12 (Institute of Medicine and Food and Nutrition Board 1998)

Group	RDA (μg day^{-1})
1-3 years	0.9
4-8 years	1.2
9-13 years	1.8
14-18 years	2.4
19-70 years	2.4
> 70 years	2.4
Pregnant women	2.6
Lactating women	2.8

The Estimated Average Requirement (EAR) consists in the value of daily intake that satisfies the vitamin B12 requirement in half of the individuals in a determined life-stage or group. The Recommended Nutrient Intake (RNI) for vitamin B12, is calculated as the EAR value plus 2 standard deviations (SD). The Food and Agriculture Organization of the United Nations (FAO)/World Health Organization (WHO) adopted the same EAR values proposed by the IOM. Moreover, the RNI values presented are equivalent to RDA values presented by the IOM (FAO and

WHO 2004). The EAR and RNI values for each group are listed in Table 11.

Table 11. Estimated average requirement (EAR) and Recommended Nutrient Intake (RNI) for vitamin B12 (FAO and WHO 2004)

Group	EAR (μg day^{-1})	RNI (μg day^{-1})
0-6 months	0.3	0.4
7-12 months	0.6	0.7
1-3 years	0.7	0.9
4-6 years	1.0	1.2
7-9 years	1.5	1.8
10-18 years	2.0	2.4
19-65 years	2.0	2.4
> 65 years	2.0	2.4
Pregnant women	2.2	2.6
Lactating women	2.4	2.8

Table 12. Vitamin B12 adequate intake (AI) for vitamin B12 (EFSA panel on dietetic products, nutrition, and allergies 2015)

Group	AI (μg day^{-1})
7-11 months	1.5
1-3 years	1.5
4-6 years	1.5
7-10 years	2.5
11-14 years	3.5
15-17 years	4.0
> 18 years	4.0
Pregnant women	4.5
Lactating women	5.0

The European Food Safety Authority (EFSA) Panel on Dietetic Products, Nutrition and Allergies (NDA) derived Dietary Reference Values (DRVs) for cobalamin, considering the available data related to cobalamin intake and its status, functions and health consequences. It was concluded that the uncertainties with respect to the cut-off values of the indicators for cobalamin insufficiency did not allow to establish an Average Requirement (AR), being an Adequate Intake (AI) for cobalamin defined instead (EFSA

Panel on Dietetic Products, Nutrition, and Allergies 2015). The AI values are summarized in Table 12.

3.9. Cellular Absorption

Vitamin B12 is a water-soluble vitamin, with a transport mechanism being necessary for passing through the cellular membrane, process in which three binding proteins, intrinsic factor, haptocorrin, and transcobalamin II, are involved (Banerjee 2006).

The absorption of vitamin B12 begins when the vitamin is released from food and reaches the stomach, were it binds to haptocorrin, produced in the salivary glands. This protein protects vitamin B12 from acid degradation owing to the stomach low pH. Once in the duodenum, a change in the pH value occurs, haptocorrin is degraded and the vitamin binds to an intrinsic factor, produced in parietal cells. This complex endocytosis into enterocytes is mediated by cubam receptor. Once inside lysosomes, the intrinsic factor is degraded and vitamin B12 is released. The free vitamin leaves the enterocyte by multidrug resistance protein 1 (MDR1), being transported to the portal circulation, where it circulates complexed to transcobalamin II, its transporter. The internalized vitamin is essentially stored in the liver and a part is excreted in bile (Green et al. 2017; Brito et al. 2018; Banerjee 2006).

3.10. Vitamin Deficiency

Vitamin B12 deficiency is manifested at several levels. The consequences may occur at haematological level, including macrocytic or megaloblastic anaemia, leukopenia, pancytopenia, thrombocytopenia and thrombocytosis, or at neurological level, such as peripheral neuropathy and areflexia. In cases of a more severe deficiency, dementia diseases and psychosis may occur (Shipton and Thachil 2015; Langan and Goodbred 2017).

The deficiency in vitamin B12 may have many causes, being pernicious anaemia one of them, since in such condition there is a destruction of the parietal cells, in which the intrinsic factor is produced. The low intake of vitamin B12 due to a poor diet can also lead to deficiency. Currently, food-bound cobalamin malabsorption (FBCM) is the most frequent cause, in which there is a difficulty in releasing the vitamin from transport proteins (Shipton and Thachil 2015).

Deficiency can be diagnosed by quantification of some biomarkers, namely, total serum vitamin B12, homocysteine, methylmalonic acid (MMA) and total serum holo-transcobalamin. Methylmalonic acid is the most reliable marker for determination of the B12 status in all age groups with normal renal function. In a vitamin B12 deficiency scenario, (R)-methylmalonyl-CoA mutase is inactive, leading to an increase of MMA, which enters blood circulation in a free form (Hannibal et al. 2016).

Vitamin B12 deficiency is very common in elderly people, affecting about 5% of the population aged 65 to 74 years and over 10% of those aged above 75 years (Clarke et al. 2004). As age increases, there is a higher risk of developing atrophic gastritis with a decrease in acid pepsin production, which is important for the cleavage of protein-bound vitamin B12 from the natural chemical form of vitamin B12. Thereby, it is recommended, for people aged over 50, the consumption of foods fortified with crystalline B12 or B12-containing supplements (Park and Johnson 2006; Baik and Russell 1999).

In vegetarians, vitamin B12 deficiency is a common occurrence, since the consumption of animal foods is very low or absent, and the vitamin presence in plant foods is very scarce. Within this group, vegans have the highest risk, as they do not ingest any food of animal origin (Rizzo et al. 2016; Pawlak, Lester, and Babatunde 2014).

Data from a systematic review of the literature reporting the cobalamin serum concentration among people following vegetarian diets, showed that 0% to 86.5% of adults and elderly, approximately 45% of infants, 0% to 33.3% of children and adolescents, and 17% to 39% of pregnant women presented vitamin B12 deficiency (Pawlak, Lester, and Babatunde 2014). In order to prevent this, dietary/pharmacological solutions are recommended, such as the consumption of supplements and fortified products (Mądry et al. 2012).

3.11. Food Sources

3.11.1. Animal Origin Products

Vitamin B12 is produced by bacteria and archaea, while plants and animals do not possess such capacity. However, after being synthetized, vitamin B12 accumulates in animal tissues, thus being a good source. In some cases, vitamin B12 may occur in plants and mushrooms through microbial interactions. Humans acquire vitamin B12 produced by bacteria mainly through the consumption of meat, specially ruminants, and fish (Watanabe and Bito 2018).

The consumption of meat, milk and fish is associated with the increase of vitamin B12 content in serum (Brouwer-Brolsma et al. 2015).

The meat consumed by humans can be derived from ruminant animals, such as beef, veal and lamb, or from omnivorous animals, namely pork and poultry (Brouwer-Brolsma et al. 2015). Ruminants' diet is essentially constituted by plants, which have no vitamin B12 content. However, they possess vitamin B12 synthetizing bacteria inhabiting their stomachs. The vitamin produced in the stomach is absorbed in the intestine, transferred to blood and stored in liver and muscles, or secreted into milk (Ortigues-Marty et al. 2005). Furthermore, omnivores include animals in their diet, which are vitamin B12 sources. The vitamin B12 content is higher in ruminant meats than in omnivorous meats. Although in milk the content is not high, particularly in comparison to meat, it is an important source since milk or dairy products are highly consumed by populations (Watanabe and Bito 2018). The vitamin B12 content of some of animal derived products is presented in Table 13.

Fish and shellfish are also a good source of vitamin B12. In aquatic environment, this vitamin can be taken by phytoplankton, that is preyed by zooplankton. Zooplankton is consumed by small fish, that serve as feeding to the bigger predatory fish, where the content of vitamin B12 is high

(Watanabe and Bito 2018). Shellfish, namely edible bivalves and snails, also have a relevant content of vitamin B12 (Watanabe and Tanioka 2014).

Table 13. Vitamin B12 content in animal-derived foods (Watanabe and Bito 2018)

Product	Vitamin B12 (μg $100g^{-1}$)
Cow Raw Liver	52.8
Cow Raw Meat	1.0-2.0
Pork Raw Liver	25.2
Pork Raw Meat	0.5
Chicken Raw Liver	44.4
Chicken Raw Meat	<0.5
Cow Milk	0.35
Sheep Milk	0.71
Goat Milk	0.06
Round Herring Viscera	37.5
Round Herring Meat	12.2
Edible Bivalves	60
Edible Snails	20

3.11.2. Fermented Products

Several vegetable origin fermented foods possess vitamin B12 in their nutritional content. Tempeh, an Indonesia traditional soybean fermented food, is an example with an high content of vitamin B12, which can vary from 0.7–8 μg $100g^{-1}$ (Nout and Rombouts 1990). In a study performed by Liem and colleagues (1977) it was verified that a Toronto tempeh sample contained 63 ng g^{-1} of vitamin B12, while the lowest value was detected in an Indonesian sample, with a 4 ng g^{-1} content. It was also observed that the inoculation of bacterium, isolated and purified from the Toronto tempeh, into sterile soybean substrate increased the vitamin B12 content to 148 ng g^{-1}.

A study of Korean centenary diets showed that traditional soybean-fermented foods such as Doenjang, Chungkookjang, Gochujang and Ganjang (soy sauce) contained a considerable amount of vitamin B12, which was higher in traditional homemade products than in the commercial versions. The vitamin B12 detected varied between 0.07–9.82 μg $100g^{-1}$

dry-weight (DW) in Doenjang, 0.05–1.40 μg 100 g^{-1} DW in Chungkookjang, 0.00–0.43 μg 100 g^{-1} DW in Gochujang and 0.02–6.76 μg 100 mL^{-1} in Ganjang (soy sauce). The vitamin B12 content in salt-fermented fish sauce made from shrimp (0.79–0.91 μg 100 mL^{-1}) or anchovy (1.52–1.77 μg 100 mL^{-1}) was lower comparatively to soy sauce. Vitamin B12 was also detected in a vegetable-fermented food named Kimchi (0.18–0.24 μg 100 g^{-1} DW) (Kwak et al. 2010).

In Asia, teas are divided according to their manufacturing processes. Half-fermented tea is referred to as blue tea, fully fermented black tea is referred to as red tea and fermented-black tea (by bacteria) is designated black tea. Small amounts of vitamin B12 can be found in several fermented tea leaves. In a study performed by Kittaka-Katsura, Watanabe, and Nakano (2004), the vitamin B12 content was determined by chemiluminescence and a microbiologic method and, according to the methods, the contents determined in blue tea were 0.068-0.081 and 0.525-0.528 μg 100 g^{-1} DW, in red tea were 0.061 and 0.663 μg 100 g^{-1} DW and in black tea were 0.104-0.859 and 0.305-1.20 μg 100g^{-1} DW, respectively. Higher values were obtained by the microbiological method, which can be due to the presence of inactive corrinoids (Kittaka-Katsura et al. 2004; Kittaka-Katsura, Watanabe, and Nakano 2004). In one other study performed by the same authors (Kittaka-Katsura et al. 2004), using Japanese fermented black tea (Batabata-cha), it was demonstrated the presence of a considerable amount of vitamin B12, reaching values of 456 ± 39 ng 100 g^{-1} (chemiluminescence) and 368 ± 56 ng 100 g^{-1} (microbiological method) in dry tea leaves and 2.0 ± 0.3 ng 100 mL^{-1} (chemiluminescence) and 2.0 ± 0.8 ng 100 mL^{-1} (microbiological method) in tea drink. The HPLC analysis strongly suggested that the compound purified from the fermented tea leaves was an active form of vitamin B12, bioavailable for mammals. This was corroborated by *in vivo* studies, that showed that vitamin B12 deficient 20-week-old rats fed with 50 mL day^{-1} of tea drink (1 ng of vitamin B12) for 6 weeks significantly decreased the excretion of methylmalonic acid in urine (an index of vitamin B12 deficiency), in comparison with the deficient rats.

Lactic acid bacteria fermentation can enhance the vitamin B12 content of vegetable-origin products. It was also verified through a microbiological assay that *P. pentosaceus* fermentation can lead to the enrichment with B12 (12.5 µg 100 mL^{-1}) of juice of fenugreek (*Trigonella foenum graecum*) leaves, that otherwise had no vitamin B12 content. A fermented soy-yogurt with vitamin B12 was produced by the fermentation of *L. reuteri* strain ZJ03 in soymilk. The vitamin B12 content was determined through microbiological assay. When the fermentation media was supplemented only with 20 g L^{-1} of glucose, a content of 68.10 ± 1.28 ng mL^{-1} was achieved. When the media was supplemented with 20 g L^{-1} of glucose, 5 g L^{-1} of glycerol and 1 g L^{-1} or 5 g L^{-1} of fructose, an 84.78 ± 4.89 or 82.19 ± 2.55 ng mL^{-1} content, respectively, was achieved. Finally, when the media was supplemented with 20 g L^{-1} of glucose, 10 g L^{-1} of glycerol and 1 g L^{-1} of fructose, the highest content of 180.30 ± 5.87 ng mL^{-1} was obtained. Considering that the content in this soy-yogurt product is higher than in other fermented soybean-based food, it can be used as an alternative food for preventing deficiency (Gu et al. 2015). The effects of a novel soymilk beverage containing 20 µg L^{-1} of a compound with B12 activity, produced by *L. reuteri* CRL1098, to prevent the pathologies caused by a B12-deficient diet were evaluated using an experimental murine model. It was observed that all symptoms relying on a nutritional B12 deficiency, both in female mice and in their offspring, were prevented, which indicates that soybean-based functional food has potential for preventing vitamin B12 deficiency. In a recent study, the fermentation of cauliflower-white bean mixture (50:50) with four *L. plantarum* strains (299v, Lp900, 299 and Heal19) was investigated. Of the assessed strains, only *L. plantarum* strain 299 (origin in human gastrointestinal tract) significantly improved the content of vitamin B12, being verified a 66% increase, reaching a value of 0.048 ± 0.013 µg 100 g^{-1} (Thompson et al. 2020).

Besides LAB, supplementation can be performed using propionic acid bacteria, namely, food grade *P. freudenreichii*, since it can synthetize active vitamin B12 forms. The production of vitamin B12 in durum flour, whole-wheat flour and wheat bran after *P. freudenreichii* subsp. *freudenreichii* DSM 20271 was added, achieved 33 ± 4, 87 ± 10 and 155 ±

17 ng g^{-1} DW, respectively, being the last two matrices the most promising for vitamin B12 *in situ* production (Xie et al. 2018). However, this production method can favour the growth of potential pathogens during fermentation. The co-fermentation of *P. freudenreichii* and *L. brevis* detain the potential to solve this issue, improving microbial safety of wheat bran (Xie et al. 2019).

Soybeans are one of the most important plant-based protein sources. Due to its high demand, it is important to find alternatives and white lupin beans could be used to overcome this dependency in several products, such as in the previously mentioned Indonesian tempeh. Moreover, it has already been shown that vitamin B12 in tempeh is not produced by fungus *Rhizopus oligosporus*, responsible for tempeh fermentation (Keuth and Bisping 1993), but by contaminating bacteria, such as *Klebsiella pneumoniae* and *Citrobacter freundii* (Keuth and Bisping 1994). As such, it was verified that adding *P. freudenreichi* subsp. *freudenreichii* DSM 20271 enhanced lupin tempeh vitamin B12 content to 0.97 µg 100g^{-1} of fresh tempeh, which is a promising result concerning *in situ* vitamin B12 fortification (Wolkers-Rooijackers, Endika, and Smid 2018). In another study, the possibility of increasing the vitamin B12 content in lupin tempeh by co-fermentation of *R. oligosporus* and *P. freudenreichii* strain 184 was also assessed, and a content of 1230 ± 58 ng g^{-1} DW of vitamin B12 was achieved after 72 h of co-fermentation. The amount obtained was 66% higher than the content achieved in sequential fermentation (684 ± 46 ng g^{-1} DW). Therefore, it was concluded that the vitamin production was enhanced by the synergistic effects of these microorganisms in co-fermentation (Signorini et al. 2018).

In situ production of active vitamin B12 by three *P. freudenreichii* strains (256, 263 and 266) in cereal matrices has also been exploited, employing different flour proportion matrices (33% in barley malt matrix, 6% in the barley flour matrix, and 15% in the wheat aleurone matrix). A B12 content of 12–37 µg kg^{-1} was synthetized in barley malt matrix and a value of 9–10 µg kg^{-1} was achieved in barley flour matrix and wheat aleurone matrix. Cobalt supplementation can lead to a 10-fold increase in the vitamin B12 content. Since DMBI cannot be used in food production, supplementation with riboflavin and nicotinamide may be an alternative. The B12 synthesis in cobalt-supplemented barley malt matrix after riboflavin and nicotinamide addition increased to 716 μg kg^{-1} and in wheat aleurone matrix, a 1.6-fold vitamin B12 increase was achieved with strain 266. Such results indicate that those compounds are possible food grade substitutes for DMBI (Chamlagain et al. 2018).

The search for novel products with enhanced nutritional value has been increasing and new products are being tested. Guava is a tropical fruit that is widely grown in tropical and subtropical regions and can be consumed fresh or as processed juice, soft drinks, jellies and baby food. Guava pulp can be used for formulation of novel fermented products and it was developed an edible gummy jelly from guava pulp fermented with *L. plantarum* WU-P19. The final product contained 1.07 mg 100 g^{-1} of vitamin B12, which means that only 1 g of guava gummy jelly could provide 10.7 μg of vitamin B12 (Palachum et al. 2020).

All fermented foods mentioned and their vitamin B12 content are summarized in Table 14.

Other methodologies can be adopted to develop vitamin B12 enriched products. Vitamin B12 fortified yogurts were created by adding isolated or microencapsulated methylcobalamin or cyanocobalamin in a concentration of 50 μg of B12 *per* 175 g of yogurt. Cyanocobalamin was shown to be the most stable form throughout storage, in fortified yogurts (Melo et al. 2020).

3.12. Vitamin B12 Biological Activity in Food

Some factors can affect vitamin B12 biological activity in food. Many bacteria synthesize pseudocobalamin, a cobalamin analogue in which the DMBI ligand was replaced by adenine (Helliwell et al. 2016). One such example is the LAB *L. reuteri* CRL1098, which is a pseudocobalamin producer (Santos et al. 2007). Pseudocobalamin is considered as unavailable for mammals since intrinsic factor, a vitamin B12 binding protein, has low affinity to pseudocobalamin (Stupperich and Nexø 1991). Thus, it is necessary to determine which analogue is present in enriched foods or supplements to ensure/assure its efficacy (Helliwell et al. 2016).

Table 14. Vitamin B12 content in fermented foods

Biofortified Foods	Micoorganism	Vitamin B12 content	Reference
Indonesia tempeh	Endogenous	0.7–8 μg 100g-1	Nout and Rombouts, 1990
Toronto tempeh – With bacteria inoculation	Endogenous	63 ng g-1	Liem et al. 1977
Toronto tempeh – Without bacteria inoculation	Bacteria isolated from Toronto tempeh	148 ng g-1	
Doenjang	Endogenous	0.07–9.82 μg 100g^{-1} DW	Kwak et al. 2010
Chungkookjang		0.05–1.40 μg 100g^{-1} DW	
Gochujang		0.00–0.43 μg 100g^{-1} DW	
Ganjang (soy sauce)		0.02–6.76 μg 100mL^{-1}	
Shrimp fish sauce		0.78–0.91 μg 100mL^{-1}	
Anchovy fish sauce		1.52–1.77 μg 100mL^{-1}	
Kimchi		0.18–0.24 μg 100g^{-1} DW	
Blue tea		0.525–0.528 μg 100g^{-1} DW	Kittaka-Katsura, Watanabe, and Nakano 2004
Red tea		0.663 μg 100g^{-1} DW	
Black tea		0.305–1.200 μg 100g^{-1} DW	
Japanese fermented black tea – Dry leaves		368 ± 56 ng 100g^{-1}	Kittaka-Katsura et al. 2004
Japanese fermented black tea – Tea		2.0 ± 0.8 ng 100 mL^{-1}	
Juice of fenugreek	*P. pentosaceus*	12.5 μg 100mL^{-1}	Gupta et al. 1998
Fermented soy-yogurt	*L. reuteri* ZJ03	68.10 ± 1.28 – 180.30 ± 5.87 ng mL^{-1}	Gu et al. 2015
Cauliflower-white bean mixture	*L. plantarum* 299	0.048 ± 0.013 μg 100g^{-1}	Thompson et al. 2020
Durum flour	*P. freudenreichii* subsp. *freudenreichii* DSM 20271	33 ng g^{-1} DW	Xie et al. 2018
Wholewheat flour		87 ng g^{-1} DW	
Wheat bran		155 ng g^{-1} DW	
Lupin tempeh		0.97 μg 100g^{-1} DW	Wolkers-Rooijackers et al. 2018
Lupin tempeh	*P. freudenreichii* 184	684–1230 ng g^{-1} DW	Signorini et al. 2018
Barley malt	Dairy originated *P. freudenreichii* 256, 263 and 266	12–37 μg kg^{-1}	Chamlagain et al. 2018
Barley flour		9–10 μg kg^{-1}	
Wheat aleurone			
Guava gummy jelly	*L. plantarum* WU-P19	1.07 mg 100g^{-1}	Palachum et al. 2020

Cooking and food processing can also affect biological activity of vitamin B12. For instance, round herring's meats vitamin B12 content significantly decreased up to approximately 62% during cooking processes, which included grilling, boiling, frying, steaming, and microwaving (Nishioka et al. 2011). Microwave heating has a significant impact in vitamin B12 stability, and heating hydroxocobalamin, the predominant form in foods, results in the generation of degradation products (Watanabe et al. 1998).

Chemical reactions and light exposure can interfere with vitamin B12 integrity. It was verified that the simultaneous presence of vitamin C and cooper can lead to the destruction of vitamin B12, originating degradation products. However, degradation can be significantly reduced by the addition of antioxidants such as carnosine or anserine (Takenaka et al. 1997). Studies performed in milk, when exposed to fluorescent light, showed a decrease ranging from 1% to 27% in cobalamin concentration (Watanabe et al. 2000).

3.13. Conclusion

Vitamin B12 represents a cobalt corrinoid family, particularly the cobalamin group. It is widely found in nature, since it can be synthetized by several microorganisms, either via *de novo* or salvage pathway.

It is produced commercially through fermentation processes using selected and genetically modified microorganisms. *P. shermanii* and *Ps. denitrificans* are frequently used in industrial production due to their capability of producing high levels of vitamin B12. Most of the research has focused on optimizing fermentation processes, mainly through traditional methods. However, there is a need to develop metabolic engineering technics, either to be applied to industrially important strains which cannot produce vitamin B12, or to strains that produce it at low level, to enhance the production.

Lactic acid bacteria are used as starter cultures in food fermentations, adding overall quality to the products. After the finding of *L. reuteri* CRL

1098 capability of growing in a vitamin B12 free medium, several LAB have been described as cobalamin producers, having a great potential of being used in formulations of fortified foods.

Vitamin B12 has a biological role in humans, being essential to the activity of two enzymes. Its deficiency, caused by health conditions, such as pernicious anaemia, or a dietary low intake, is manifested at several levels, having haematological and neurological consequences.

Vegetarians are a risk group due their diet's composition, which does not comprise animal products, which are the main source of vitamin B12. Fermented vegetable products are interesting alternatives to supply vitamin B12. Therefore, strategies have been developed to increase vitamin B12 content in fermented products, such as adding vitamin B12 synthetizing bacteria to the fermentation process. It should be noticed that in addition to the vitamin B12 content in food products it is also important to evaluate its biological activity.

Acknowledgments

This work was supported by project ALGAVALOR - Microalgas: Integrated Production and Valuation of Biomass and Miscellaneous Applications (POCI-01-0247-FEDER-035234; LISBOA-01-0247-FEDER-035234; ALG-01-0247-FEDER-035234) funded by Lisboa 2020; ALGARVE 2020; COMPETE 2020; Portugal 2020; FEDER - Fundo Europeu de Desenvolvimento Regional; AICEP - Agência para o Investimento e Comércio Externo de Portugal, E.P.E. The authors would also like to thank the scientific collaboration under the FCT project UID/Multi/50016/2019 and FCT fellowship SFRH/BD/104016/2014.

References

Abbas, Charles A, and Andriy A Sibirny. 2011. "Genetic Control of Biosynthesis and Transport of Riboflavin and Flavin Nucleotides and

Construction of Robust Biotechnological Producers." *Microbiology and Molecular Biology Reviews : MMBR* 75 (2): 321–60. https://doi.org/10.1128/MMBR.00030-10.

Acevedo-Rocha, Carlos G, Luisa S Gronenberg, Matthias Mack, Fabian M Commichau, and Hans J Genee. 2019. "Microbial Cell Factories for the Sustainable Manufacturing of B Vitamins." *Current Opinion in Biotechnology* 56: 18–29. https://doi.org/https://doi.org/10.1016/j.copbio.2018.07.006.

Aguiar, Tatiana Q, Rui Silva, and Lucília Domingues. 2015. "Ashbya Gossypii beyond Industrial Riboflavin Production: A Historical Perspective and Emerging Biotechnological Applications." *Biotechnology Advances* 33 (8): 1774–86. https://doi.org/10.1016/j.biotechadv.2015.10.001.

Ahmad, Iqbal, Wahid Hussain, and Azhar Ali Fareedi. 1992. "Photolysis of Cyanocobalamin in Aqueous Solution." *Journal of Pharmaceutical and Biomedical Analysis* 10 (1): 9–15. https://doi.org/https://doi.org/10.1016/0731-7085(92)80004-7.

Akiyama, T, J Selhub, and I H Rosenberg. 1982. "FMN Phosphatase and FAD Pyrophosphatase in Rat Intestinal Brush Borders: Role in Intestinal Absorption of Dietary Riboflavin." *The Journal of Nutrition* 112 (2): 263–68. https://doi.org/10.1093/jn/112.2.263.

Albano, Clara, Tiziana Silvetti, and Milena Brasca. 2020. "Screening of Lactic Acid Bacteria Producing Folate and Their Potential Use as Adjunct Cultures for Cheese Bio-Enrichment." *FEMS Microbiology Letters* 367 (9). https://doi.org/10.1093/femsle/fnaa059.

Allen, Robert H, Sally P Stabler, and John Lindenbaum. 1993. "Serum Betaine, N,N-Dimethylglycine and N-Methylglycine Levels in Patients with Cobalamin and Folate Deficiency and Related Inborn Errors of Metabolism." *Metabolism* 42 (11): 1448–60. https://doi.org/https://doi.org/10.1016/0026-0495(93)90198-W.

Angelis, Maria De, Francesca Bottacini, Bruno Fosso, Philip Kelleher, Maria Calasso, Raffaella Di Cagno, Marco Ventura, Ernesto Picardi, Douwe van Sinderen, and Marco Gobbetti. 2014. "Lactobacillus Rossiae, a Vitamin B12 Producer, Represents a Metabolically

Versatile Species within the Genus Lactobacillus." *PloS One* 9 (9): e107232. https://doi.org/10.1371/journal.pone.0107232.

Angier, R B, J H Boothe, B L Hutchings, J H Mowat, J Semb, E L Stokstad, Y Subbarow, et al. 1945. "Synthesis of a Compound Identical with the *L. casei* Factor Isolated from Liver." *Science (New York, N.Y.)* 102 (2644): 227–28. https://doi. org/10.1126/science. 102.2644.227.

Ashoori, Marziyeh, and Ahmad Saedisomeolia. 2014. "Riboflavin (Vitamin B2) and Oxidative Stress: A Review." *British Journal of Nutrition* 111 (11): 1985–91. https://doi.org/10.1017/S0007114514 000178.

Auclair, Olivia, Yang Han, and Sergio A. Burgos. 2019. "Consumption of Milk and Alternatives and Their Contribution to Nutrient Intakes among Canadian Adults: Evidence from the 2015 Canadian Community Health Survey-Nutrition." *Nutrients* 11 (8): 1–17. https://doi.org/10.3390/nu11081948.

Bacher, A., S. Eberhardt, and G. Richter. 1996. "Biosynthesis of Riboflavin." In *Escherichia Coli and Salmonella: Cellular and Molecular Biology*, edited by F. C. Neidhardt, R. Curtiss III, J. L. Ingraham, E. C. C. Lin, K. B. Low, B. Magasanik, W. S. Reznikoff, M. Riley, M. Schaechter, and H. E. Umbarger, 2nd ed., 657–664. Washington, D.C.: ASM Press.

Bacher, A, S Eberhardt, M Fischer, K Kis, and G Richter. 2000. "Biosynthesis of Vitamin B2 (Riboflavin)." *Annual Review of Nutrition* 20: 153–67. https://doi.org/10.1146/annurev.nutr.20.1.153.

Baik, H W, and R M Russell. 1999. "Vitamin B12 Deficiency in the Elderly." *Annu Rev Nutr* 19: 357–77. https://doi.org/10.1146/ annurev.nutr.19.1.357.

Bailey, Matthew R, and Zachary D Schultz. 2016. "SERS Speciation of the Electrochemical Oxidation-Reduction of Riboflavin." *The Analyst* 141 (17): 5078–87. https://doi.org/10.1039/c6an01054g.

Balasubramaniam, Shanti, John Christodoulou, and Shamima Rahman. 2019. "Disorders of Riboflavin Metabolism." *Journal of Inherited Metabolic Disease* 42 (4): 608–19. https://doi.org/10.1002/jimd.12058.

Ball, G F M. 2005. *Vitamins in Foods: Analysis, Bioavailability, and Stability*. CRC Press.

Banerjee, R. 2006. "B12 Trafficking in Mammals: A for Coenzyme Escort Service." *ACS Chem Biol* 1 (3): 149–59. https://doi.org/10.1021/cb6001174.

Banerjee, R V, and R G Matthews. 1990. "Cobalamin-Dependent Methionine Synthase." *FASEB Journal : Official Publication of the Federation of American Societies for Experimental Biology* 4 (5): 1450–59. https://doi.org/10.1096/fasebj.4.5.2407589.

Barker, H A, H Weissbach, and R D Smyth. 1958. "A Coenzyme Containing Pseudovitamin B(12)." *Proceedings of the National Academy of Sciences of the United States of America* 44 (11): 1093–97. https://doi.org/10.1073/pnas.44.11.1093.

Basavanna, G, and S G Prapulla. 2013. "Evaluation of Functional Aspects of Lactobacillus Fermentum CFR 2195 Isolated from Breast Fed Healthy Infants' Fecal Matter." *J Food Sci Technol* 50 (2): 360–66. https://doi.org/10.1007/s13197-011-0345-9.

Bates, C J, and N J Fuller. 1986. "The Effect of Riboflavin Deficiency on Methylenetetrahydrofolate Reductase (NADPH)(EC 1.5.1.20) and Folate Metabolism in the Rat." *British Journal of Nutrition* 55 (2): 455–64. https://doi.org/DOI: 10.1079/BJN19860051.

Beinert, H. 1960. "Flavin Coenzymes." In *The Enzymes*, edited by P. Boyer, H. Lardy, and K. Myrback, Vol. 2, 339–416. New York: Academic Press.

Berezovskii, V. M. 1973. *Chemistry of Vitamins*. Edited by V. M. Berezovskii. Moscow, USSR: Food Industry Publishing House.

Berry Ottaway, P. 2010. *Stability of Vitamins during Food Processing and Storage. Chemical Deterioration and Physical Instability of Food and Beverages*. Woodhead Publishing Limited. https://doi.org/10.1533/9781845699260.3.539.

Bhushan, Bharat, C R Kumkum, Mamta Kumari, Jayesh J Ahire, Leon M T Dicks, and Vijendra Mishra. 2020. "Soymilk Bio-Enrichment by Indigenously Isolated Riboflavin-Producing Strains of Lactobacillus

Plantarum." *Lwt* 119 (November 2019). https://doi.org/10.1016/j.lwt.2019.108871.

Bhushan, Bharat, S K Tomar, and Surajit Mandal. 2016. "Phenotypic and Genotypic Screening of Human-Originated Lactobacilli for Vitamin B12 Production Potential: Process Validation by Micro-Assay and UFLC." *Applied Microbiology and Biotechnology* 100 (15): 6791–6803. https://doi.org/10.1007/s00253-016-7639-9.

Blitz, Milton, Edward Eigen, and Ephraim Gunsberg. 1956. "Studies Relating to The Stability of Vitamin B12 in B-Complex Injectable Solutions**Analytical and Control Laboratories, U. S. Vitamin Corp., Yonkers, N. Y." *Journal of the American Pharmaceutical Association (Scientific Ed.)* 45 (12): 803–6. https://doi.org/https://doi.org/10.1002/jps.3030451210.

Blount, Kenneth F, and Ronald R Breaker. 2006. "Riboswitches as Antibacterial Drug Targets." *Nature Biotechnology* 24 (12): 1558–64. https://doi.org/10.1038/nbt1268.

Blyth. A W. 1879. "No Title." *J. Chem. SOC.* 35: 530–39.

Bonnett, R. 1963. "The Chemistry of the Vitamin B12 Group." *Chemical Reviews* 63 (6): 573–605. https://doi.org/10.1021/cr60226a002.

Bonnett, R, J R Cannon, V M Clark, A W Johnson, L F J Parker, E Lester Smith, and Alexander Todd. 1957. "227. Chemistry of the Vitamin B12 Group. Part V. The Structure of the Chromophoric Grouping." *Journal of the Chemical Society (Resumed)*, no. 0: 1158–68. https://doi.org/10.1039/JR9570001158.

Brito, Alex, Edwin Habeych, Irma Silva-Zolezzi, Nicola Galaffu, and Lindsay H Allen. 2018. "Methods to Assess Vitamin B12 Bioavailability and Technologies to Enhance Its Absorption." *Nutrition Reviews* 76 (10): 778–92. https://doi.org/10.1093/nutrit/nuy026.

Brouwer-Brolsma, Elske M, Rosalie A M Dhonukshe-Rutten, Janneke P van Wijngaarden, Nikita L van der Zwaluw, Nathalie van der Velde, and Lisette C P G M de Groot. 2015. "Dietary Sources of Vitamin B-12 and Their Association with Vitamin B-12 Status Markers in

Healthy Older Adults in the B-PROOF Study." *Nutrients* 7 (9): 7781–97. https://doi.org/10.3390/nu7095364.

Burch, R. 2011. *The Stability and Shelf Life of Vitamin-Fortified Foods. Food and Beverage Stability and Shelf Life*. Woodhead Publishing Limited. https://doi.org/10.1533/9780857092540.3.743.

Burgess, Catherine M, Eddy J Smid, Ger Rutten, and Douwe van Sinderen. 2006. "A General Method for Selection of Riboflavin-Overproducing Food Grade Micro-Organisms." *Microbial Cell Factories* 5 (1): 24. https://doi.org/10.1186/1475-2859-5-24.

Burgess, Catherine, Mary O'Connell-Motherway, Wilbert Sybesma, Jeroen Hugenholtz, and Douwe Van Sinderen. 2004. "Riboflavin Production in Lactococcus Lactis: Potential for In Situ Production of Vitamin-Enriched Foods." *Applied and Environmental Microbiology* 70 (10): 5769–77. https://doi.org/10.1128/AEM.70.10.5769.

Bykhovskiĭ, V Ia, N I Zaĭtseva, and A A Eliseev. 1998. "[Tetrapyrroles: diversity, biosynthesis, biotechnology]." *Prikladnaia biokhimiia i mikrobiologiia* 34 (1): 3–21.

Camacho, L, C Sierra, D Marcus, E Guzmán, R Campos, D von Bäer, and L Trugo. 1991. "Nutritional Quality of Lupine (Lupinus Albus Cv. Multolupa) as Affected by Lactic Acid Fermentation." *International Journal of Food Microbiology* 14 (3–4): 277–86. https://doi.org/10.1016/0168-1605(91)90119-A.

Capozzi, Vittorio, Valeria Menga, Anna Maria Digesù, Pasquale De Vita, Douwe Van Sinderen, Luigi Cattivelli, Clara Fares, and Giuseppe Spano. 2011. "Biotechnological Production of Vitamin B2-Enriched Bread and Pasta." *Journal of Agricultural and Food Chemistry* 59 (14): 8013–20. https://doi.org/10.1021/jf201519h.

Capozzi, Vittorio, Pasquale Russo, María Teresa Dueñas, Paloma López, and Giuseppe Spano. 2012. "Lactic Acid Bacteria Producing B-Group Vitamins: A Great Potential for Functional Cereals Products." *Applied Microbiology and Biotechnology* 96 (6): 1383–94. https://doi.org/10.1007/s00253-012-4440-2.

Cárdenas, Nivia, Jonathan E. Laiño, Susana Delgado, Esther Jiménez, Marianela Juárez del Valle, Graciela Savoy de Giori, Fernando Sesma, et al. 2015. "Relationships between the Genome and Some Phenotypical Properties of Lactobacillus Fermentum CECT 5716, a Probiotic Strain Isolated from Human Milk." *Applied Microbiology and Biotechnology* 99 (10): 4343–53. https://doi.org/10.1007/s00253-015-6429-0.

Cardoso, Daniel R, Silvia H Libardi, and Leif H Skibsted. 2012. "Riboflavin as a Photosensitizer. Effects on Human Health and Food Quality." *Food and Function* 3 (5): 487–502. https://doi.org/10.1039/c2fo10246c.

Carrizo, Silvana L, Cecilia E Montes de Oca, Jonathan E Laiño, Nadia E Suarez, Graciela Vignolo, Jean Guy LeBlanc, and Graciela Rollán. 2016. "Ancestral Andean Grain Quinoa as Source of Lactic Acid Bacteria Capable to Degrade Phytate and Produce B-Group Vitamins." *Food Research International* 89: 488–94. https://doi.org/10.1016/j.foodres.2016.08.013.

Carrizo, Silvana L, Cecilia E Montes De Oca, María Elvira Hébert, Lucila Saavedra, Graciela Vignolo, Jean Guy LeBlanc, and Graciela Celestina Rollán. 2017. "Lactic Acid Bacteria from Andean Grain Amaranth: A Source of Vitamins and Functional Value Enzymes." *Journal of Molecular Microbiology and Biotechnology* 27 (5): 289–98. https://doi.org/10.1159/000480542.

Carrizo, Silvana L, Alejandra de Moreno de LeBlanc, Jean Guy LeBlanc, and Graciela C Rollán. 2020. "Quinoa Pasta Fermented with Lactic Acid Bacteria Prevents Nutritional Deficiencies in Mice." *Food Research International* 127: 108735. https://doi.org/https://doi.org/10.1016/j.foodres.2019.108735.

Chamlagain, Bhawani, Tessa A Sugito, Paulina Deptula, Minnamari Edelmann, Susanna Kariluoto, Pekka Varmanen, and Vieno Piironen. 2018. "In Situ Production of Active Vitamin B12 in Cereal Matrices Using Propionibacterium Freudenreichii." *Food Science & Nutrition* 6 (1): 67–76. https://doi.org/10.1002/fsn3.528.

Chen, Jun, Mike Vestergaard, Thomas Glasdam Jensen, Jing Shen, Martin Dufva, Christian Solem, and Peter Ruhdal Jensena. 2017. "Finding the Needle in the Haystack-the Use of Microfluidic Droplet Technology to

Identify Vitamin-Secreting Lactic Acid Bacteria." *MBio* 8 (3): 1–12. https://doi.org/10.1128/mBio.00526-17.

Choe, Eunok, Rongmin Huang, and D B Min. 2005. "*Chemical Reactions and Stability of Riboflavin in Foods*" 70 (1): 1–10. https://doi.org/10.1111/j.1365-2621.2005.tb09055.x.

Clarke, Robert, J Grimley Evans, J Schneede, E Nexo, C Bates, A Fletcher, A Prentice, et al. 2004. "Vitamin B12 and Folate Deficiency in Later Life." *Age and Ageing* 33 (1): 34–41. https://doi.org/10.1093/ageing/afg109.

Crécy-Lagard, Valérie de, Basma El Yacoubi, Rocío Díaz de la Garza, Alexandre Noiriel, and Andrew D Hanson. 2007. "Comparative Genomics of Bacterial and Plant Folate Synthesis and Salvage: Predictions and Validations." *BMC Genomics* 8 (1): 245. https://doi.org/10.1186/1471-2164-8-245.

Curti, B, S Ronchi, and P M Simonetta. 1992. "D- and l-Amino Acid Oxidases." In *Chemistry and Biochemistry of Flavoenzyme*, edited by F. Müller, Vol. III, 69–94. Boca Raton: CRC Press.

Daniel, H, E Binninger, and G Rehner. 1983. "Hydrolysis of FMN and FAD by Alkaline Phosphatase of the Intestinal Brush-Border Membrane." *International Journal for Vitamin and Nutrition Research. Internationale Zeitschrift Fur Vitamin- Und Ernahrungsforschung. Journal International de Vitaminologie et de Nutrition* 53 (1): 109–14.

Demerre, Leon J, and Catherine Wilson. 1956. "Photolysis of Vitamin B12." *Journal of the American Pharmaceutical Association (Scientific Ed.)* 45 (3): 129–34. https://doi.org/https://doi.org/10.1002/jps.3030450302.

Deponte, Marcel. 2013. "Glutathione Catalysis and the Reaction Mechanisms of Glutathione-Dependent Enzymes." *Biochimica et Biophysica Acta* 1830 (5): 3217–66. https://doi.org/10.1016/j.bbagen.2012.09.018.

DiBaise, Michelle, and Sherry M Tarleton. 2019. "Hair, Nails, and Skin: Differentiating Cutaneous Manifestations of Micronutrient

Deficiency." *Nutrition in Clinical Practice* 34 (4): 490–503. https://doi.org/10.1002/ncp.10321.

Edwards, A L, and R T Batey. 2010. "Riboswitches: A Common RNA Regulatory Element." *Nature Education* 3 (9): 9.

EFSA. 2017. "Dietary Reference Values for Riboflavin." *EFSA Journal* 15 (8). https://doi.org/10.2903/j.efsa.2017.4919.

EFSA Panel on Dietetic Products, Nutrition, and Allergies, EFSA. 2015. "Scientific Opinion on Dietary Reference Values for Cobalamin (Vitamin B12)." *EFSA Journal* 13 (7): 1–64. https://doi.org/10.2903/j.efsa.2015.4150.

EFSA Panel on Dietetic Products, Nutrition and Alergies (NDA), EFSA. 2014. "Scientific Opinion on Dietary Reference Values for Folate." *EFSA Journal* 12 (11): 1–59. https://doi.org/10.2903/j.efsa.2014.3893.

Eitenmiller, Ronald R, W O Landen Jr., and Lin Ye. 2007. *Vitamin Analysis for the Health and Food Sciences*. Boca Raton: CRC Press. https://doi.org/https://doi.org/10.1201/9781420009750.

Elizaquível, Patricia, Alba Pérez-Cataluña, Alba Yépez, Cecilia Aristimuño, Eugenia Jiménez, Pier Sandro Cocconcelli, Graciela Vignolo, and Rosa Aznar. 2015. "Pyrosequencing vs. Culture-Dependent Approaches to Analyze Lactic Acid Bacteria Associated to Chicha, a Traditional Maize-Based Fermented Beverage from Northwestern Argentina." *International Journal of Food Microbiology* 198: 9–18. https://doi.org/10.1016/j.ijfoodmicro.2014.12.027.

Escalante-Semerena, Jorge C. 2007. "Conversion of Cobinamide into Adenosylcobamide in Bacteria and Archaea." *Journal of Bacteriology* 189 (13): 4555–60. https://doi.org/10.1128/JB.00503-07.

Fang, Huan, Jie Kang, and Dawei Zhang. 2017. "Microbial Production of Vitamin B12: A Review and Future Perspectives." *Microbial Cell Factories* 16 (1): 1–14. https://doi.org/10.1186/s12934-017-0631-y.

Fang, Huan, Dong Li, Jie Kang, Pingtao Jiang, Jibin Sun, and Dawei Zhang. 2018. "Metabolic Engineering of Escherichia Coli for de Novo Biosynthesis of Vitamin B(12)." *Nature Communications* 9 (1): 4917. https://doi.org/10.1038/s41467-018-07412-6.

FAO, and WHO. 2004. "Vitamin and Mineral Requirements in Human Nutrition Second Edition." *World Health Organization*, 1–20. https://doi.org/92 4 154612 3.

Feller, Beate A, and Thomas J Macek. 1955. "Effect of Thiamine Hydrochloride on the Stability of Solutions of Crystalline Vitamin B12." *Journal of the American Pharmaceutical Association (Scientific Ed.)* 44 (11): 662–65. https://doi.org/https://doi.org/10.1002/jps.3030441106.

Fitzpatrick, Teresa B, Gilles J C Basset, Patrick Borel, Fernando Carrari, Dean DellaPenna, Paul D Fraser, Hanjo Hellmann, et al. 2012. "Vitamin Deficiencies in Humans: Can Plant Science Help?" *Plant Cell* 24 (2): 395–414. https://doi.org/10.1105/tpc.111.093120.

Flatz, G, 1970. "Enhanced Binding of FAD to Glutathione Reductase in G6PD Deficiency." *Nature* 226 (5247): 755. https://doi.org/10.1038/226755a0.

Foraker, Amy B, Chandra M Khantwal, and Peter W Swaan. 2003. "Current Perspectives on the Cellular Uptake and Trafficking of Riboflavin." *Advanced Drug Delivery Reviews* 55 (11): 1467–83. https://doi.org/10.1016/j.addr.2003.07.005.

García-Angulo, Víctor Antonio. 2017. "Overlapping Riboflavin Supply Pathways in Bacteria." *Critical Reviews in Microbiology* 43 (2): 196–209. https://doi.org/10.1080/1040841X.2016.1192578.

Garst, Andrew D, and Robert T Batey. 2009. "A Switch in Time: Detailing the Life of a Riboswitch." *Biochimica et Biophysica Acta - Gene Regulatory Mechanisms* 1789 (9–10): 584–91. https://doi.org/10.1016/j.bbagrm.2009.06.004.

Ge, Ying-Ying, Jia-Rong Zhang, Harold Corke, and Ren-You Gan. 2020. "Screening and Spontaneous Mutation of Pickle-Derived Lactobacillus Plantarum with Overproduction of Riboflavin, Related Mechanism, and Food Application." *Foods* 9 (88): 1–12. https://doi.org/10.3390/foods9010088.

Gibson, R S, 2005. Assessment of the Status of Thiamin, Riboflavin, and Niacin. Edited by R.S. Gibson. *Principles of Nutritional Assessment*. 2nd ed. New York: Oxford University Press.

Green, Ralph, Lindsay H Allen, Anne-Lise Bjørke-Monsen, Alex Brito, Jean-Louis Guéant, Joshua W Miller, Anne M Molloy, et al. 2017. "Vitamin B(12) Deficiency." *Nature Reviews. Disease Primers* 3 (June): 17040. https://doi.org/10.1038/nrdp.2017.40.

Gropper, S S, J L Smith, and J L Groff. 2009. "Riboflavin." In *Advanced Nutrition and Human Metabolism*, 329–33. Wadsworth: CENGAG Learning.

Gu, Qing, Chen Zhang, Dafeng Song, Ping Li, and Xuan Zhu. 2015. "Enhancing Vitamin B12 Content in Soy-Yogurt by Lactobacillus Reuteri." *International Journal of Food Microbiology* 206 (August): 56–59. https://doi.org/10.1016/j.ijfoodmicro.2015.04.033.

Guest, J R, S Friedman, D D Woods, and E L Smith. 1962. "A Methyl Analogue of Cobamide Coenzyme in Relation to Methionine Synthesis by Bacteria." *Nature* 195 (July): 340–42. https://doi.org/10.1038/195340a0.

Gupta, U, Rudramma, E R Rati, and R Joseph. 1998. "Nutritional Quality of Lactic Fermented Bitter Gourd and Fenugreek Leaves." *International Journal of Food Sciences and Nutrition* 49 (2): 101–8. https://doi.org/10.3109/09637489809089389.

Guru V, and Viswanathan K. 2013. "Riboflavin Production in Milk Whey Using Probiotic Bacteria-Lactobacillus Acidophilus and Lactococcus Lactis." *Indian Journal of Fundamental and Applied Life Sciences* 3 (4): 169–76. http://www.cibtech.org/jls.htm.

Guyot, Jean Pierre. 2012. "Cereal-Based Fermented Foods in Developing Countries: Ancient Foods for Modern Research." *International Journal of Food Science and Technology* 47 (6): 1109–14. https://doi.org/10.1111/j.1365-2621.2012.02969.x.

Haase, Ilka, Tobias Gräwert, Boris Illarionov, Adelbert Bacher, and Markus Fischer. 2014. "Recent Advances in Riboflavin Biosynthesis." *Methods in Molecular Biology (Clifton, N.J.)* 1146: 15–40. https://doi.org/10.1007/978-1-4939-0452-5_2.

Hannibal, Luciana, Vegard Lysne, Anne-Lise Bjørke-Monsen, Sidney Behringer, Sarah C Grünert, Ute Spiekerkoetter, Donald W Jacobsen, and Henk J Blom. 2016. "Biomarkers and Algorithms for the

Diagnosis of Vitamin B12 Deficiency." *Frontiers in Molecular Biosciences* 3: 27. https://doi.org/10.3389/fmolb.2016.00027.

Hati, Subrota, Maulik Patel, Birendra K. Mishra, and Sujit Das. 2019. "Short-Chain Fatty Acid and Vitamin Production Potentials of Lactobacillus Isolated from Fermented Foods of Khasi Tribes, Meghalaya, India." *Annals of Microbiology* 69 (11): 1191–99. https://doi.org/10.1007/s13213-019-01500-8.

Helliwell, Katherine Emma, Andrew David Lawrence, Andre Holzer, Ulrich Johan Kudahl, Severin Sasso, Bernhard Kräutler, David John Scanlan, Martin James Warren, and Alison Gail Smith. 2016. "Cyanobacteria and Eukaryotic Algae Use Different Chemical Variants of Vitamin B12." *Current Biology : CB* 26 (8): 999–1008. https://doi.org/10.1016/j.cub.2016.02.041.

Hemery, Youna M, Laura Fontan, Arnaud Laillou, Vincent Jallier, Regina Moench-Pfanner, Sylvie Avallone, and Jacques Berger. 2020. "Influence of Storage Conditions and Packaging of Fortified Wheat Flour on Microbial Load and Stability of Folate and Vitamin B12." *Food Chemistry: X* 5: 100076. https://doi.org/https://doi.org/10.1016/j.fochx.2019.100076.

Hemmerich, P, C Veeger, and H C Wood. 1965. "Progress in the Chemistry and Molecular Biology of Flavins and Flavocoenzymes." *Angewandte Chemie (International Ed. in English)* 4 (8): 671–87. https://doi.org/10.1002/anie.196506711.

Hill, M J, 1997. "Intestinal Flora and Endogenous Vitamin Synthesis." *European Journal of Cancer Prevention.* https://doi.org/10.1097/00008469-199703001-00009.

Hodgkin, Dorothy Crowfoot, Jennifer Kamper, Maureen Mackay, and Jennifer Pickworth. 1956. "The Structure of Vitamin B12." *Nature* 178: 64–66. https://doi.org/https://doi.org/10.1038/178064a0-8.

Hodgkin, Dorothy Crowfoot, Jenny Pickworth, John H Robertson, Kenneth N Trueblood, Richard J Prosen, and John G White. 1955. "Structure of Vitamin B12:The Crystal Structure of the Hexacarboxylic Acid Derived from B12 and the Molecular Structure of the Vitamin." *Nature* 176 (4477): 325–28. https://doi.org/10.1038/176325a0.

Hodgkin, Dorothy, Jennifer Kamper, June Lindsey, Maureen Mackay, Jenny Pickworth, J H Robertson, Clara Brink Shoemaker, J G White, R J Prosen, and N Trueblood. 1957. "The Structure of Vitamin B 12 . I. An Outline of the Crystallographic Investigation of Vitamin B 12." *Proceedings of the Royal Society of London. Series A. Mathematical and Physical Sciences* 242 (1229): 228–63. https://doi.org/10.1098/rspa.1957.0174.

Hoffbrand, A V, and D G, Weir. 2001. "The History of Folic Acid." *British Journal of Haematology* 113: 579–89. https://doi.org/10.1080/00268976.2013.846029.

Hüfner, Eric, Robert A Britton, Stefan Roos, Hans Jonsson, and Christian Hertel. 2008. "Global Transcriptional Response of Lactobacillus Reuteri to the Sourdough Environment." *Systematic and Applied Microbiology* 31 (5): 323–38. https://doi.org/10.1016/j.syapm.2008.06.005.

Hümbelin, M, V Griesser, T Keller, W Schurter, M Haiker, H P Hohmann, H Ritz, G Richter, A Bacher, and A P G M van Loon. 1999. "GTP Cyclohydrolase II and 3,4-Dihydroxy-2-Butanone 4-Phosphate Synthase Are Rate-Limiting Enzymes in Riboflavin Synthesis of an Industrial Bacillus Subtilis Strain Used for Riboflavin Production." *Journal of Industrial Microbiology and Biotechnology* 22 (1): 1–7. https://doi.org/10.1038/sj.jim.2900590.

Institute of Medicine, and Food and Nutrition Board. 1998. *Dietary Reference Intakes for Thiamin, Riboflavin, Niacin, Vitamin B6, Folate, Vitamin B12, Pantothenic Acid, Biotin and Choline: A Report of the Standing Committee on the Scientific Evaluation of Dietary Reference Intakes and Its Panel on Folate, Other B. Trends in Food Science & Technology*. https://doi.org/10.1016/S0924-2244(01)00010-3.

Jacques, Paul F, Renee Kalmbach, Pamela J Bagley, Giuseppina T Russo, Gail Rogers, Peter W F Wilson, Irwin H Rosenberg, and Jacob Selhub. 2002. "The Relationship between Riboflavin and Plasma Total Homocysteine in the Framingham Offspring Cohort Is Influenced by Folate Status and the C677T Transition in the Methylenetetrahydrofolate Reductase Gene." *The Journal of Nutrition* 132 (2): 283–88. https://doi.org/10.1093/jn/132.2.283.

Jägerstad, M, and K Arkbåge. 2003. "COBALAMINS | Properties and Determination." In *Encyclopedia of Food Sciences and Nutrition*, edited by B. Caballero, Second Edi. Academic Press: Oxford.

Jayashree, S, J Rajendhran, K Jayaraman, G Kalaichelvan, and P. Gunasekaran. 2011. "Improvement of Riboflavin Production by Lactobacillus Fermentum Isolated from Yogurt." *Food Biotechnology* 25 (3): 240–51. https://doi.org/10.1080/08905436.2011.590769.

Jayashree, Sathyanarayanan, Kunthala Jayaraman, and Gurumurthy Kalaichelvan. 2010. "Isolation, Screening and Characterization of Riboflavin." *Recent Research in Science and Technology* 2 (1): 83–88.

Jiao, Wenshu, Song Wang, Jiaqi Guan, Jialu Shi, Smith Etareri Evivie, Fenfen Yan, Na Li, Junliang Chen, Bailiang Li, and Guicheng Huo. 2020. "Milk Fermented with Lactococcus Lactis KLDS4.0325 Alleviates Folate Status in Deficient Mice." *Food & Function* 11 (5): 4571–81. https://doi.org/10.1039/c9fo03067k.

Jiménez, Alberto, María A Santos, Markus Pompejus, and José L Revuelta. 2005. "Metabolic Engineering of the Purine Pathway for Riboflavin Production in Ashbya Gossypii." *Applied and Environmental Microbiology* 71 (10): 5743–51. https://doi.org/10.1128/AEM.71.10.5743-5751.2005.

Jiménez, Eugenia, Alba Yépez, Alba Pérez-Cataluña, Elena Ramos Vásquez, Doris Zúñiga Dávila, Graciela Vignolo, and Rosa Aznar. 2018. "Exploring Diversity and Biotechnological Potential of Lactic Acid Bacteria from Tocosh - Traditional Peruvian Fermented Potatoes - by High Throughput Sequencing (HTS) and Culturing." *LWT - Food Science and Technology* 87: 567–74. https://doi.org/10.1016/j.lwt.2017.09.033.

Juarez Del Valle, M, J E Laiño, A de Moreno de LeBlanc, G Savoy de Giori, and J G LeBlanc. 2016. "Soyamilk Fermented with Riboflavin-Producing Lactobacillus Plantarum CRL 2130 Reverts and Prevents Ariboflavinosis in Murine Models." *The British Journal of Nutrition* 116 (7): 1229–35. https://doi.org/10.1017/S0007114516003378.

Juarez del Valle, M, J E Laiño, G Savoy de Giori, and J G LeBlanc. 2014. "Riboflavin Producing Lactic Acid Bacteria as a Biotechnological Strategy to Obtain Bio-Enriched Soymilk." *Food Research International* 62: 1015–19. https://doi.org/https://doi.org/10.1016/j.foodres.2014.05.029.

Juzeniene, Asta, and Zivile Nizauskaite. 2013. "Photodegradation of Cobalamins in Aqueous Solutions and in Human Blood." *Journal of Photochemistry and Photobiology B: Biology* 122: 7–14. https://doi.org/https://doi.org/10.1016/j.jphotobiol.2013.03.001.

Kang, Zhen, Junli Zhang, Jingwen Zhou, Qingsheng Qi, Guocheng Du, and Jian Chen. 2012. "Recent Advances in Microbial Production of δ-Aminolevulinic Acid and Vitamin B12." *Biotechnology Advances* 30 (6): 1533–42. https://doi.org/https://doi.org/10.1016/j.biotechadv.2012.04.003.

Kaprasob, Ratchadaporn, Orapin Kerdchoechuen, Natta Laohakunjit, and Promluck Somboonpanyakul. 2018. "B Vitamins and Prebiotic Fructooligosaccharides of Cashew Apple Fermented with Probiotic Strains *Lactobacillus* Spp., Leuconostoc Mesenteroides and Bifidobacterium Longum." *Process Biochemistry* 70: 9–19. https://doi.org/https://doi.org/10.1016/j.procbio.2018.04.009.

Kariluoto, Susanna, Minnamari Edelmann, Laura Nyström, Tuula Sontag-Strohm, Hannu Salovaara, Reetta Kivelä, Mirkka Herranen, Matti Korhola, and Vieno Piironen. 2014. "In Situ Enrichment of Folate by Microorganisms in Beta-Glucan Rich Oat and Barley Matrices." *International Journal of Food Microbiology* 176 (April): 38–48. https://doi.org/10.1016/j.ijfoodmicro.2014.01.018.

Karrer, P, K Schöpp, and F Benz. 1935. "Synthesen von Flavinen IV." *Helvetica Chimica Acta* 18 (1): 426–29. https://doi.org/doi:10.1002/hlca.19350180152.

Kehoe, L, J Walton, S M Hopkins, B A McNulty, A P Nugent, H McNulty, M Ward, and A Flynn. 2018. "Intake, Status and Dietary Sources of Riboflavin in a Representative Sample of Irish Adults Aged 18–90 Years." *Proceedings of the Nutrition Society* 77 (OCE3): 4919. https://doi.org/10.1017/s0029665118000708.

Keuth, S, and B Bisping. 1994. "Vitamin B12 Production by Citrobacter Freundii or Klebsiella Pneumoniae during Tempeh Fermentation and Proof of Enterotoxin Absence by PCR." *Applied and Environmental Microbiology* 60 (5): 1495–99.

Keuth, Sylvia, and B Bisping. 1993. "Formation of Vitamins by Pure Cultures of Tempe Moulds and Bacteria during the Tempe Solid Substrate Fermentation." *Journal of Applied Bacteriology* 75 (5): 427–34. https://doi.org/10.1111/j.1365-2672.1993.tb02798.x.

Khalili, Mohammad, Aziz Homayouni Rad, Ahmad Yari Khosroushahi, Hadi Khosravi, and Somayeh Jafarzadeh. 2020. "Application of Probiotics in Folate Bio-Fortification of Yoghurt." *Probiotics and Antimicrobial Proteins* 12 (2): 756–63. https://doi.org/10.1007/s12602-019-09560-7.

Kim, J J, M Wang, and R Paschke. 1993. "Crystal Structures of Medium-Chain Acyl-CoA Dehydrogenase from Pig Liver Mitochondria with and without Substrate." *Proceedings of the National Academy of Sciences of the United States of America* 90 (16): 7523–27. https://doi.org/10.1073/pnas.90.16.7523.

Kim, Kyung Hyun, Byung Hee Chun, Ju Hye Baek, Seong Woon Roh, Se Hee Lee, and Che Ok Jeon. 2020. "Genomic and Metabolic Features of Lactobacillus Sakei as Revealed by Its Pan-Genome and the Metatranscriptome of Kimchi Fermentation." *Food Microbiology* 86 (April): 103341. https://doi.org/10.1016/j.fm.2019.103341.

Kittaka-Katsura, Hiromi, Syuhei Ebara, Fumio Watanabe, and Yoshihisa Nakano. 2004. "Characterization of Corrinoid Compounds from a Japanese Black Tea (Batabata-Cha) Fermented by Bacteria." *Journal of Agricultural and Food Chemistry* 52 (4): 909–11. https://doi.org/10.1021/jf030585r.

Kittaka-Katsura, Hiromi, Fumio Watanabe, and Yoshihisa Nakano. 2004. "Occurrence of Vitamin B12 in Green, Blue, Red, and Black Tea Leaves." *Journal of Nutritional Science and Vitaminology* 50 (6): 438–40. https://doi.org/10.3177/jnsv.50.438.

Kleerebezem, Michiel, Jos Boekhorst, Richard van Kranenburg, Douwe Molenaar, Oscar P Kuipers, Rob Leer, Renato Tarchini, et al. 2003.

"Complete Genome Sequence of Lactobacillus Plantarum WCFS1." *Proceedings of the National Academy of Sciences* 100 (4): 1990–95. https://doi.org/10.1073/pnas.0337704100.

Ko, Yeounjoo, Somasundar Ashok, Satish Kumar Ainala, Mugesh Sankaranarayanan, Ah Yeong Chun, Gyoo Yeol Jung, and Sunghoon Park. 2014. "Coenzyme B12 Can Be Produced by Engineered Escherichia Coli under Both Anaerobic and Aerobic Conditions." *Biotechnology Journal* 9 (12): 1526–35. https://doi.org/10.1002/biot.201400221.

Kodentsova, V M, O A Vrzhesinskaia, E V Trofimenko, A A Sokol'nikov, N A Beketova, N V Blazheevich, V A Isaeva, S I Aleĭnik, L S Trofimenko, and V I Dronova. 1994. "[Vitamin status of children with diabetes mellitus]." *Voprosy meditsinskoi khimii* 40 (2): 45–48.

Korhola, M, R Hakonen, K Juuti, M Edelmann, S Kariluoto, L Nyström, T Sontag-Strohm, and V Piironen. 2014. "Production of Folate in Oat Bran Fermentation by Yeasts Isolated from Barley and Diverse Foods." *Journal of Applied Microbiology* 117 (3): 679–89. https://doi.org/10.1111/jam.12564.

Krajewski, Stefanie Sandra, Dmitry Ignatov, and Jörgen Johansson. 2017. "Two Are Better Than One: Dual Targeting of Riboswitches by Metabolite Analogs." *Cell Chemical Biology* 24 (5): 535–37. https://doi.org/10.1016/j.chembiol.2017.05.004.

Kuhn, Richard. 1936. "Lactoflavin (Vitamin B2)." *Angewandte Chemie* 49 (1): 6–10. https://doi.org/doi:10.1002/ange.19360490103.

Kuhn, Richard, K Reinemund, and F Weygand. 1934. "Synthesis of Lumi-Lactoflavin." *Dtsh. Chem. Ges.* 67: 1460–63.

Kwak, Chung Shil, Mee Sook Lee, Se In Oh, and Sang Chul Park. 2010. "Discovery of Novel Sources of Vitamin b(12) in Traditional Korean Foods from Nutritional Surveys of Centenarians." *Current Gerontology and Geriatrics Research* 2010: 374897. https://doi.org/10.1155/2010/374897.

Laiño, Jonathan Emiliano, Marianela Juarez del Valle, Graciela Savoy de Giori, and Jean Guy Joseph LeBlanc. 2013. "Development of a High Folate Concentration Yogurt Naturally Bio-Enriched Using Selected

Lactic Acid Bacteria." *LWT - Food Science and Technology* 54 (1): 1–5. https://doi.org/https://doi.org/10.1016/j.lwt.2013.05.035.

Laiño, Jonathan Emiliano, Jean Guy Leblanc, and Graciela Savoy de Giori. 2012. "Production of Natural Folates by Lactic Acid Bacteria Starter Cultures Isolated from Artisanal Argentinean Yogurts." *Canadian Journal of Microbiology* 58 (5): 581–88. https://doi.org/10.1139/w2012-026.

Laird, E, and A M Molly. 2014. "Water-Soluble Vitamins and Essential Nutrients." *Reference Module in Biomedical Sciences (Elsevier).*

Lakhan, Ram, Veedamali S Subramanian, and Hamid M Said. 2017. "Role of MicroRNA-423-5p in Posttranscriptional Regulation of the Intestinal Riboflavin Transporter-3." *American Journal of Physiology. Gastrointestinal and Liver Physiology* 313 (6): G589–98. https://doi.org/10.1152/ajpgi.00238.2017.

Langan, Robert C, and Andrew J Goodbred. 2017. "Vitamin B12 Deficiency: Recognition and Management." *American Family Physician* 96 (6): 384–89.

LeBlanc, J G, J E Laiño, M Juarez del Valle, V Vannini, D van Sinderen, M P Taranto, G Font de Valdez, G Savoy de Giori, and F Sesma. 2011. "B-Group Vitamin Production by Lactic Acid Bacteria--Current Knowledge and Potential Applications." *Journal of Applied Microbiology* 111 (6): 1297–1309. https://doi.org/10.1111/j.1365-2672.2011.05157.x.

LeBlanc, Jean Guy, Jonathan Emiliano Laiño, Marianela Juárez del Valle, Graciela Savoy de Giori, Fernando Sesma, and María Pía Taranto. 2015. "B-Group Vitamins Production by Probiotic Lactic Acid Bacteria." *Biotechnology of Lactic Acid Bacteria: Novel Applications: Second Edition*, 279–96. https://doi.org/10.1002/9781118868386.ch17.

LeBlanc, Jean Guy, Romina Levit, Graciela Savoy de Giori, and Alejandra de Moreno de LeBlanc. 2020. "Application of Vitamin-Producing Lactic Acid Bacteria to Treat Intestinal Inflammatory Diseases." *Applied Microbiology and Biotechnology* 104 (8): 3331–37. https://doi.org/10.1007/s00253-020-10487-1.

Lee, Elaine R, Kenneth F Blount, and Ronald R Breaker. 2009. "Roseoflavin Is a Natural Antibacterial Compound That Binds to FMN Riboswitches and Regulates Gene Expression." *RNA Biology* 6 (2): 187–94. https://doi.org/10.4161/rna.6.2.7727.

Lenhert, P G, and D C Hodgkin. 1961. "Structure of the 5,6-Dimethyl-Benzimidazolylcobamide Coenzyme." *Nature* 192 (December): 937–38. https://doi.org/10.1038/192937a0.

Levit, R, G Savoy de Giori, A de Moreno de LeBlanc, and J G LeBlanc. 2017. "Evaluation of the Effect of Soymilk Fermented by a Riboflavin-Producing Lactobacillus Plantarum Strain in a Murine Model of Colitis." *Beneficial Microbes* 8 (1): 65–72. https://doi.org/10.3920/BM2016.0063.

Levit, Romina, Graciela Savoy de Giori, Alejandra de Moreno de LeBlanc, and Jean Guy LeBlanc. 2019. "Beneficial Effect of a Mixture of Vitamin-Producing and Immune-Modulating Lactic Acid Bacteria as Adjuvant for Therapy in a Recurrent Mouse Colitis Model." *Applied Microbiology and Biotechnology* 103 (21–22): 8937–45. https://doi.org/10.1007/s00253-019-10133-5.

Li, Jing, Yong Ge, Mojgan Zadeh, Roy 3rd Curtiss, and Mansour Mohamadzadeh. 2020. "Regulating Vitamin B12 Biosynthesis via the CbiMCbl Riboswitch in Propionibacterium Strain UF1." *Proceedings of the National Academy of Sciences of the United States of America* 117 (1): 602–9. https://doi.org/10.1073/pnas.1916576116.

Li, Ping, Qing Gu, Yuejiao Wang, Yue Yu, Lanlan Yang, and Jieyan V Chen. 2017. "Novel Vitamin B(12)-Producing Enterococcus Spp. and Preliminary in Vitro Evaluation of Probiotic Potentials." *Applied Microbiology and Biotechnology* 101 (15): 6155–64. https://doi.org/10.1007/s00253-017-8373-7.

Li, Ping, Qingqing Zhou, and Qing Gu. 2016. "Complete Genome Sequence of Lactobacillus Plantarum LZ227, a Potential Probiotic Strain Producing B-Group Vitamins." *Journal of Biotechnology* 234: 66–70. https://doi.org/10.1016/j.jbiotec.2016.07.020.

Li, Shan Shan, Yi Wei Xu, Jian Yi Wu, Hua Zhen Tan, Zhi Yong Wu, Yu Jie Xue, Jian Jun Zhang, En Min Li, and Li Yan Xu. 2017. "Plasma

Riboflavin Level Is Associated with Risk, Relapse, and Survival of Esophageal Squamous Cell Carcinoma." *Nutrition and Cancer* 69 (1): 21–28. https://doi.org/10.1080/01635581.2017.1247890.

Liem, I T, K H Steinkraus, and T C Cronk. 1977. "Production of Vitamin B-12 in Tempeh, a Fermented Soybean Food." *Applied and Environmental Microbiology* 34 (6): 773–76.

Lienhart, Wolf-Dieter, Venugopal Gudipati, and Peter Macheroux. 2013. "The Human Flavoproteome." *Archives of Biochemistry and Biophysics* 535 (2): 150–62. https://doi.org/10.1016/j.abb.2013.02.015.

Liu, Dandan, and Janos Zempleni. 2014. "Low Activity of LSD1 Elicits a Pro-Inflammatory Gene Expression Profile in Riboflavin-Deficient Human T Lymphoma Jurkat Cells." *Genes & Nutrition* 9 (5): 422. https://doi.org/10.1007/s12263-014-0422-6.

Liu, Shuang, Wenya Hu, Zhiwen Wang, and Tao Chen. 2020. "Production of Riboflavin and Related Cofactors by Biotechnological Processes." *Microbial Cell Factories* 19 (1): 31. https://doi.org/10.1186/s12934-020-01302-7.

Madhu, Arenahalli Ningegowda, Prashant Giribhattanavar, Mandayam Singara Narayan, and Siddalingaiya Gurudutt Prapulla. 2010. "Probiotic Lactic Acid Bacterium from Kanjika as a Potential Source of Vitamin B12: Evidence from LC-MS, Immunological and Microbiological Techniques." *Biotechnology Letters* 32 (4): 503–6. https://doi.org/10.1007/s10529-009-0176-1.

Mądry, Edyta, Aleksandra Lisowska, Philip Grebowiec, and Jarosław Walkowiak. 2012. "The Impact of Vegan Diet on B-12 Status in Healthy Omnivores: Five-Year Prospective Study." *Acta Scientiarum Polonorum. Technologia Alimentaria* 11 (2): 209–12.

Magnúsdóttir, Stefanía, Dmitry Ravcheev, Valérie De Crécy-Lagard, and Ines Thiele. 2015. "Systematic Genome Assessment of B-Vitamin Biosynthesis Suggests Cooperation among Gut Microbes." *Frontiers in Genetics* 6 (MAR). https://doi.org/10.3389/fgene.2015.00148.

Mandal, Maumita, and Ronald R Breaker. 2004. "Gene Regulation by Riboswitches." *Nature Reviews. Molecular Cell Biology* 5 (6): 451–63. https://doi.org/10.1038/nrm1403.

Mande, Sharmila S, Steve Sarfaty, Mark D Allen, Richard N Perham, and Wim G J Hol. 1996. "Protein-Protein Interactions in the Pyruvate Dehydrogenase Multienzyme Complex: Dihydrolipoamide Dehydrogenase Complexed with the Binding Domain of Dihydrolipoamide Acetyltransferase." *Structure* 4 (3): 277–86. https://doi.org/10.1016/S0969-2126(96)00032-9.

Marashly, Eyad T, and Saeed A Bohlega. 2017. "Riboflavin Has Neuroprotective Potential: Focus on Parkinson's Disease and Migraine." *Frontiers in Neurology* 8 (JUL): 1–12. https://doi.org/10.3389/fneur.2017.00333.

Martens, J H, H Barg, M J Warren, and D Jahn. 2002. "Microbial Production of Vitamin B12." *Applied Microbiology and Biotechnology* 58 (3): 275–85. https://doi.org/10.1007/s00253-001-0902-7.

Martín, R, M Olivares, M L Marín, J Xaus, L Fernández, and J M Rodríguez. 2005. "Characterization of a Reuterin-Producing Lactobacillus Coryniformis Strain Isolated from a Goat's Milk Cheese." *International Journal of Food Microbiology* 104 (3): 267–77. https://doi.org/10.1016/j.ijfoodmicro.2005.03.007.

Martín, Rocío, Susana Langa, Carlota Reviriego, Esther Jiménez, María L. Marín, Jordi Xaus, Leonides Fernández, and Juan M. Rodríguez. 2003. "Human Milk Is a Source of Lactic Acid Bacteria for the Infant Gut." *Journal of Pediatrics* 143 (6): 754–58. https://doi.org/10.1016/j.jpeds.2003.09.028.

Massey, V. 2000. "The Chemical and Biological Versatility of Riboflavin." *Biochemical Society Transactions* 28 (4): 283–96.

Masuda, Misako, Mariko Ide, Haruka Utsumi, Tae Niiro, Yuko Shimamura, and Masatsune Murata. 2012. "Production Potency of Folate, Vitamin B(12), and Thiamine by Lactic Acid Bacteria Isolated from Japanese Pickles." *Bioscience, Biotechnology, and Biochemistry* 76 (11): 2061–67. https://doi.org/10.1271/bbb.120414.

McCormick, Donald B, 2010. "Vitamin/Mineral Supplements: Of Questionable Benefit for the General Population." *Nutrition Reviews* 68 (4): 207–13. https://doi.org/10.1111/j.1753-4887.2010.00279.x.

McKinley, Michelle C, 2000. "Nutritional Aspects and Possible Pathological Mechanisms of Hyperhomocysteinaemia: An Independent Risk Factor for Vascular Disease." *Proceedings of the Nutrition Society* 59 (2): 221–37. https://doi.org/DOI: 10.1017/S0029665100000252.

Mehdizadeh Aghdam, Elnaz, Mohammad Saeid Hejazi, and Abolfazl Barzegar. 2016. "Riboswitches: From Living Biosensors to Novel Targets of Antibiotics." *Gene* 591 (2): 244–59. https://doi.org/10.1016/j.gene.2016.07.035.

Melo, Larisse, Clementine Ng, Ricky Tsang, Anubhav Pratap Singh, David Kitts, and Yvonne Lamers. 2020. "Development of Novel Vitamin B12 Fortified Yogurts Using Isolated and Microencapsulated Vitamin B12." *Proceedings of the Nutrition Society* 79 (OCE2): 2020. https://doi.org/10.1017/s0029665120002128.

Mestdagh, F, B De Meulenaer, J De Clippeleer, F Devlieghere, and A Huyghebaert. 2005. "Protective Influence of Several Packaging Materials on Light Oxidation of Milk." *Journal of Dairy Science* 88 (2): 499–510. https://doi.org/10.3168/jds.S0022-0302(05)72712-0.

Mielgo-Ayuso, Juan, Raquel Aparicio-Ugarriza, Josune Olza, Javier Aranceta-Bartrina, Ángel Gil, Rosa M. Ortega, Lluis Serra-Majem, Gregorio Varela-Moreiras, and Marcela González-Gross. 2018. "Dietary Intake and Food Sources of Niacin, Riboflavin, Thiamin and Vitamin B6 in a Representative Sample of the Spanish Population. The Anthropometry, Intake, and Energy Balance in Spain (ANIBES) Study." *Nutrients* 10 (7). https://doi.org/10.3390/nu10070846.

Minot, George R, and William P Murphy. 1926. "Treatment of Pernicious Anemia by a Special Diet." *Journal of the American Medical Association* 87 (7): 470–76. https://doi.org/10.1001/jama.1926.02680070016005.

Monteiro, Mariana, and Daniel Perrone. 2012. "Chemistry and Biochemistry of Riboflavin and Related Compounds." In *B Vitamins and Folate : Chemistry, Analysis, Function and Effects*, edited by Victor R Preedy, 93–107. The Royal Society of Chemistry. https://doi.org/https://doi.org/10.1039/9781849734714.

Mosegaard, Signe, Graziana Dipace, Peter Bross, Jasper Carlsen, Niels Gregersen, and Rikke Katrine Jentoft Olsen. 2020. "Riboflavin Deficiency—Implications for General Human Health and Inborn Errors of Metabolism." *International Journal of Molecular Sciences* 21 (11). https://doi.org/10.3390/ijms21113847.

Mosso, A L, M E Jimenez, G Vignolo, J G LeBlanc, and N C Samman. 2018. "Increasing the Folate Content of Tuber Based Foods Using Potentially Probiotic Lactic Acid Bacteria." *Food Research International* 109: 168–74. https://doi.org/https://doi.org/10.1016/j.foodres.2018.03.073.

Mudd, S H, H L Levy, and F Skovby. 1995. "Disorders of Trans-Sulfuration." In *The Metabolic Basis of Inherited Disease*, edited by CR Scriver, A Beaudet, WS Sly, and D Valle, 7th ed., 1279–1327. New York: McGraw-Hill.

Mushtaq, Sohail, Han Su, Marilyn H E Hill, and Hilary J Powers. 2009. "Erythrocyte Pyridoxamine Phosphate Oxidase Activity: A Potential Biomarker of Riboflavin Status?" *The American Journal of Clinical Nutrition* 90 (5): 1151–59. https://doi.org/10.3945/ajcn.2009.28338.

Nahvi, Ali, Jeffrey E Barrick, and Ronald R Breaker. 2004. "Coenzyme B12 Riboswitches Are Widespread Genetic Control Elements in Prokaryotes." *Nucleic Acids Research* 32 (1): 143–50. https://doi.org/10.1093/nar/gkh167.

Nakano, Emi, Sohail Mushtaq, Paul R Heath, Eun-Sook Lee, Jonathan P Bury, Stuart A Riley, Hilary J Powers, and Bernard M Corfe. 2011. "Riboflavin Depletion Impairs Cell Proliferation in Adult Human Duodenum: Identification of Potential Effectors." *Digestive Diseases and Sciences* 56 (4): 1007–19. https://doi.org/10.1007/s10620-010-1374-3.

Nathoo, Tasnim, Christina P. Holmes, and Aleck Ostry. 2005. "An Analysis of the Development of Canadian Food Fortification Policies: The Case of Vitamin B." *Health Promotion International* 20 (4): 375–82. https://doi.org/10.1093/heapro/dai015.

Neville, Janice N, Juanita A Eagles, Gregorio Samson, and R E Olason. 1968. "Nutritional Status of Alcoholics." *The American Journal of*

Clinical Nutrition 21 (11): 1329–40. https://doi.org/10.1093/ajcn/21.11.1329.

Nishioka, Michiko, Fuki Kanosue, Yukinori Yabuta, and Fumio Watanabe. 2011. "Loss of Vitamin B(12) in Fish (Round Herring) Meats during Various Cooking Treatments." *Journal of Nutritional Science and Vitaminology* 57 (6): 432–36. https://doi.org/10.3177/jnsv.57.432.

Nout, M J R, and F M Rombouts. 1990. "Recent Developments in Tempe Research." *Journal of Applied Bacteriology* 69 (5): 609–33. https://doi.org/10.1111/j.1365-2672.1990.tb01555.x.

Nudler, Evgeny, and Alexander S. Mironov. 2004. "The Riboswitch Control of Bacterial Metabolism." *Trends in Biochemical Sciences* 29 (1): 11–17. https://doi.org/10.1016/j.tibs.2003.11.004.

Oguro, Yoshifumi, Toshikazu Nishiwaki, Ryota Shinada, Kazuya Kobayashi, and Atsushi Kurahashi. 2017. "Metabolite Profile of Koji Amazake and Its Lactic Acid Fermentation Product by Lactobacillus Sakei UONUMA." *Journal of Bioscience and Bioengineering* 124 (2): 178–83. https://doi.org/10.1016/j.jbiosc.2017.03.011.

Ortigues-Marty, Isabelle, Didier Micol, Sophie Prache, Dominique Dozias, and Christiane L Girard. 2005. "Nutritional Value of Meat: The Influence of Nutrition and Physical Activity on Vitamin B12 Concentrations in Ruminant Tissues." *Reproduction, Nutrition, Development* 45 (4): 453–67. https://doi.org/10.1051/rnd:2005038.

Otani, Shohei, Sabu Kasai, and Kunio B T - Methods in Enzymology Matsui. 1980. "[34] Isolation, Chemical Synthesis, and Properties of Roseoflavin." In *Vitamins and Coenzymes Part E*, 66:235–41. Academic Press. https://doi.org/https://doi.org/10.1016/0076-6879(80)66464-7.

Pacheco Da Silva, Fabiana Fernanda, Vanessa Biscola, Jean Guy LeBlanc, and Bernadette Dora Gombossy de Melo Franco. 2016. "Effect of Indigenous Lactic Acid Bacteria Isolated from Goat Milk and Cheeses on Folate and Riboflavin Content of Fermented Goat Milk." *LWT - Food Science and Technology* 71: 155–61. https://doi.org/10.1016/j.lwt.2016.03.033.

Palachum, Wilawan, Wanna Choorit, Supranee Manurakchinakorn, and Yusuf Chisti. 2020. "Guava Pulp Fermentation and Processing to a Vitamin B12-Enriched Product." *Journal of Food Processing and Preservation* 5 (May): 1–15. https://doi.org/10.1111/jfpp.14566.

Pan, Feng, Hong Jun Luo, Zhi Yong Wu, Su Zuan Chen, Xuan Wang, Shuai Xia Yu, Jia Min Wang, et al. 2020. "Decreased Plasma Riboflavin Is Associated with Poor Prognosis, Invasion, and Metastasis in Esophageal Squamous Cell Carcinoma." *European Journal of Clinical Nutrition*. https://doi.org/10.1038/s41430-020-0585-8.

Park, Sohyun, and Mary Ann Johnson. 2006. "What Is an Adequate Dose of Oral Vitamin B12 in Older People with Poor Vitamin B12 Status?" *Nutrition Reviews* 64 (8): 373–78. https://doi.org/10.1111/j.1753-4887.2006.tb00222.x.

Patel, Mulchand S, Natalia S Nemeria, William Furey, and Frank Jordan. 2014. "The Pyruvate Dehydrogenase Complexes: Structure-Based Function and Regulation." *The Journal of Biological Chemistry* 289 (24): 16615–23. https://doi.org/10.1074/jbc.R114.563148.

Pawlak, R, S E Lester, and T Babatunde. 2014. "The Prevalence of Cobalamin Deficiency among Vegetarians Assessed by Serum Vitamin B12: A Review of Literature." *European Journal of Clinical Nutrition* 68 (5): 541–48. https://doi.org/10.1038/ejcn.2014.46.

Pedrolli, Danielle Biscaro, Christian Kühm, Daniel C. Sévin, Michael P. Vockenhuber, Uwe Sauer, Beatrix Suess, and Matthias Mack. 2015. "A Dual Control Mechanism Synchronizes Riboflavin and Sulphur Metabolism in Bacillus Subtilis." *Proceedings of the National Academy of Sciences of the United States of America* 112 (45): 14054–59. https://doi.org/10.1073/pnas.1515024112.

Perkins, J B, and J Pero. 2002. "Vitamin Biosynthesis." In *Bacillus Subtilis and Its Closest Relatives from Genes to Cells*, edited by A. Sonenshein, J. Hoch, and R. Losick, 271–286. Washington, D.C: ASM Press.

Piao, Yongzhe, Pornpimon Kiatpapan, Mitsuo Yamashita, and Yoshikatsu Murooka. 2004. "Effects of Expression of HemA and HemB Genes on

Production of Porphyrin in Propionibacterium Freudenreichii." *Applied and Environmental Microbiology* 70 (12): 7561–66. https://doi.org/10.1128/AEM.70.12.7561-7566.2004.

Piao, Yongzhe, Mitsuo Yamashita, Nami Kawaraichi, Ryo Asegawa, Hisayo Ono, and Yoshikatsu Murooka. 2004. "Production of Vitamin B12 in Genetically Engineered Propionibacterium Freudenreichii." *Journal of Bioscience and Bioengineering* 98 (3): 167–73. https://doi.org/10.1016/S1389-1723(04)00261-0.

Pinto, J T, and J Zempleni. 2016. "Riboflavin." *Advances in Nutrition* 7 (5): 973–75. https://doi.org/10.3945/an.116.012716.

Powers, Hilary J. 2003. "Riboflavin (Vitamin B-2) and Health." *American Journal of Clinical Nutrition*. https://doi.org/10.1093/ajcn/77.6.1352.

Rajalakshmi, R, and K Vanaja. 1967. "Chemical and Biological Evaluation of the Effects of Fermentation on the Nutritive Value of Foods Prepared from Rice and Grams." *British Journal of Nutrition* 21: 467–73.

Raux, E, A Lanois, F Levillayer, M J Warren, E Brody, A Rambach, and C Thermes. 1996. "Salmonella Typhimurium Cobalamin (Vitamin B12) Biosynthetic Genes: Functional Studies in S. Typhimurium and Escherichia Coli." *Journal of Bacteriology* 178 (3): 753–67. https://doi.org/10.1128/jb.178.3.753-767.1996.

Raux, Evelyne, Heidi L Schubert, Jennifer M Roper, Keith S Wilson, and Martin J Warren. 1999. "Vitamin B12: Insights into Biosynthesis's Mount Improbable." *Bioorganic Chemistry* 27 (2): 100–118. https://doi.org/https://doi.org/10.1006/bioo.1998.1125.

Renz, P, 2007. "Investigations on the Biosynthesis of the 5,6-Dimethylbenz-Imidazole Moiety of Vitamin B12." In *Vitamin B12 and B12 Proteins*, edited by B. Kräutler, D. Arigoni, and B.T. Golding, 119–130. Weinheim, Germany: Wiley-VCH.

Rickes, E L, N G Brink, F R Koniuszy, T R Wood, and K Folkers. 1948. "Crystalline Vitamin B12." *Science* 107 (2781): 396–97. https://doi.org/10.1126/science.107.2781.396.

Rivlin, R S, 2001. "Riboflavin." In *Present Knowledge in Nutrition*, edited by B. A. Bowman and R. M. Russell, 8th ed., 191–98. Washington, DC: International Life Sciences Institute.

———. 2010. "Riboflavin." In *Encyclopedia of Dietary Supplements*, edited by PM Coates, JM Betz, MR Blackman, GM Cragg, M Levine, J Moss, and JD White, 2nd ed., 691–99. London and New York: Informa Healthcare.

Rivlin, R S, and J T Pinto. 2001. "Riboflavin (Vitamin B2)." In *Handbook of Vitamins*, edited by RB Rucker, JW Suttie, DB McCormick, and LJ Machlin, 3rd ed., 255–73. New York: CRC Press.

Rizzo, Gianluca, Antonio Simone Laganà, Agnese Maria Chiara Rapisarda, Gioacchina Maria Grazia La Ferrera, Massimo Buscema, Paola Rossetti, Angela Nigro, et al. 2016. "Vitamin B12 among Vegetarians: Status, Assessment and Supplementation." *Nutrients* 8 (12). https://doi.org/10.3390/nu8120767.

Rodrigo-Torres, Lidia, Alba Yépez, Rosa Aznar, and David R Arahal. 2019. "Genomic Insights Into Five Strains of Lactobacillus Plantarum With Biotechnological Potential Isolated From Chicha, a Traditional Maize-Based Fermented Beverage From Northwestern Argentina." *Frontiers in Microbiology* 10: 2232. https://doi.org/10.3389/fmicb.2019.02232.

Roe, M A, S Bell, M Oseredczuk, T Christensen, S Westenbrink, H Pakkala, K Presser, and P M Finglas. 2017. "Updated Food Composition Database for Nutrient Intake." *EFSA Supporting Publications* 10 (6): 1–34. https://doi.org/10.2903/sp.efsa.2013.en-355.

Rosenthal, W S, N F Adham, R Lopez, and J M Cooperman. 1973. "Riboflavin Deficiency in Complicated Chronic Alcoholism." *The American Journal of Clinical Nutrition* 26 (8): 858–60. https://doi.org/10.1093/ajcn/26.8.858.

Rossi, Madallena, Stefano Raimondi, Luca Costantino, and Alberto Amaretti. 2016. "Folate: Relevance of Chemical and Microbial Production." In *Industrial Biotechnology of Vitamins, Biopigments, and Antioxidants*, edited by Erick J. Vandamme and José Luis

Revuelta, 103–28. Wiley-VCH Verlag GmbH & Co. KGaA. https://doi.org/https://doi.org/10.1002/9783527681754.ch5.

Roth, J R, J G Lawrence, M Rubenfield, S Kieffer-Higgins, and G M Church. 1993. "Characterization of the Cobalamin (Vitamin B12) Biosynthetic Genes of Salmonella Typhimurium." *Journal of Bacteriology* 175 (11): 3303–16. https://doi.org/10.1128/jb.175.11.3303-3316.1993.

Russo, Pasquale, Vittorio Capozzi, Mattia Pia Arena, Giuseppina Spadaccino, María Teresa Dueñas, Paloma López, Daniela Fiocco, and Giuseppe Spano. 2014. "Riboflavin-Overproducing Strains of Lactobacillus Fermentum for Riboflavin-Enriched Bread." *Applied Microbiology and Biotechnology* 98 (8): 3691–3700. https://doi.org/10.1007/s00253-013-5484-7.

Russo, Pasquale, Nuria Peña, Maria Lucia Valeria de Chiara, Maria Luisa Amodio, Giancarlo Colelli, and Giuseppe Spano. 2015. "Probiotic Lactic Acid Bacteria for the Production of Multifunctional Fresh-Cut Cantaloupe." *Food Research International* 77: 762–72. https://doi.org/10.1016/j.foodres.2015.08.033.

Saade, Carol, and Emily O Arijaje. 2020. "17 - Fortification." In *Breakfast Cereals and How They Are Made: Raw Materials, Processing, and Production*, edited by Alicia A Perdon, Sylvia L Schonauer, and Kaisa S B T - Breakfast Cereals and How They Are Made (Third Edition) Poutanen, 343–59. Oxford: AACC International Press. https://doi.org/10.1016/B978-0-12-812043-9.00017-5.

Salvucci, Emiliano, Jean Guy LeBlanc, and Gabriela Pérez. 2016. "Technological Properties of Lactic Acid Bacteria Isolated from Raw Cereal Material." *LWT* 70: 185–91. https://doi.org/https://doi.org/10.1016/j.lwt.2016.02.043.

Santos-Mendoza, Ricardo, Elena Ramos-Vásquez, Amparo Iris Zavaleta, Doris Zúñiga-Dávila, and Carmen Velezmoro-Sánchez. 2018. "Production of Riboflavin by Lactic Acid Bacteria Isolated from the Manufacturing of 'Tunta.'" *Ecología Aplicada* 17 (2). https://doi.org/http://dx.doi.org/10.21704/rea.v17i2.1238.

Santos, Filipe, Jose L Vera, René van der Heijden, Graciela Valdez, Willem M de Vos, Fernando Sesma, and Jeroen Hugenholtz. 2008. "The Complete Coenzyme B12 Biosynthesis Gene Cluster of Lactobacillus Reuteri CRL1098." *Microbiology (Reading, England)* 154 (Pt 1): 81–93. https://doi.org/10.1099/mic.0.2007/011569-0.

Santos, Filipe, José L Vera, Pedro Lamosa, Graciela F de Valdez, Willem M de Vos, Helena Santos, Fernando Sesma, and Jeroen Hugenholtz. 2007. "Pseudovitamin B(12) Is the Corrinoid Produced by Lactobacillus Reuteri CRL1098 under Anaerobic Conditions." *FEBS Letters* 581 (25): 4865–70. https://doi.org/10.1016/j.febslet.2007.09.012.

Schiano, A N, Y Jo, D M Barbano, and M A Drake. 2019. "Does Vitamin Fortification Affect Light Oxidation in Fluid Skim Milk?" *Journal of Dairy Science* 102 (6): 4877–90. https://doi.org/10.3168/jds.2018-15594.

Schnellbaecher, Alisa, Dennis Binder, Stephanie Bellmaine, and Aline Zimmer. 2019. "Vitamins in Cell Culture Media: Stability and Stabilization Strategies." *Biotechnology and Bioengineering* 116 (6): 1537–55. https://doi.org/10.1002/bit.26942.

Schwechheimer, Susanne Katharina, Enoch Y Park, José Luis Revuelta, Judith Becker, and Christoph Wittmann. 2016. "Biotechnology of Riboflavin." *Applied Microbiology and Biotechnology* 100 (5): 2107–19. https://doi.org/10.1007/s00253-015-7256-z.

Sebrell, W H, and R E Butler. 1939. "Riboflavin Deficiency in Man (Ariboflavinosis)." *Public Health Reports (1896-1970)* 54 (48): 2121–31. https://doi.org/10.2307/4583104.

Serganov, A, and E Nudler. 2013. "A Decade of Riboswitches." *Cell* 152 (1): 17–24. https://doi.org/doi:10.1016/j.cell.2012.12.024. A.

Serganov, Alexander, Lili Huang, and Dinshaw J Patel. 2009. "Coenzyme Recognition and Gene Regulation by a Flavin Mononucleotide Riboswitch." *Nature* 458 (7235): 233–37. https://doi.org/10.1038/nature07642.

Serrano-Amatriain, Cristina, Rodrigo Ledesma-Amaro, Rubén López-Nicolás, Gaspar Ros, Alberto Jiménez, and José Luis Revuelta. 2016.

"Folic Acid Production by Engineered Ashbya Gossypii." *Metabolic Engineering* 38 (November): 473–82. https://doi.org/10.1016/j.ymben.2016.10.011.

Shi, Zumin, Shiqi Zhen, Gary A Wittert, Baojun Yuan, Hui Zuo, and Anne W Taylor. 2014. "Inadequate Riboflavin Intake and Anemia Risk in a Chinese Population: Five-Year Follow up of the Jiangsu Nutrition Study." *PLoS ONE* 9 (2). https://doi.org/10.1371/journal.pone.0088862.

Shipton, Michael J, and Jecko Thachil. 2015. "Vitamin B12 Deficiency - A 21st Century Perspective." *Clinical Medicine (London, England)* 15 (2): 145–50. https://doi.org/10.7861/clinmedicine.15-2-145.

Signorini, Camilla, Aristodemo Carpen, Luigi Coletto, Gigliola Borgonovo, Elisabetta Galanti, Jessica Capraro, Chiara Magni, et al. 2018. "Enhanced Vitamin B12 Production in an Innovative Lupin Tempeh Is Due to Synergic Effects of Rhizopus and Propionibacterium in Cofermentation." *International Journal of Food Sciences and Nutrition* 69 (4): 451–57. https://doi.org/10.1080/09637486.2017.1386627.

Snauwaert, Isabel, Pieter Stragier, Luc De Vuyst, and Peter Vandamme. 2015. "Comparative Genome Analysis of Pediococcus Damnosus LMG 28219, a Strain Well-Adapted to the Beer Environment." *BMC Genomics* 16 (1): 267. https://doi.org/10.1186/s12864-015-1438-z.

Stahmann, K P, H N Jr Arst, H Althöfer, J L Revuelta, N Monschau, C Schlüpen, C Gätgens, A Wiesenburg, and T Schlösser. 2001. "Riboflavin, Overproduced during Sporulation of Ashbya Gossypii, Protects Its Hyaline Spores against Ultraviolet Light." *Environmental Microbiology* 3 (9): 545–50. https://doi.org/10.1046/j.1462-2920.2001.00225.x.

Stupperich, E, and E Nexø. 1991. "Effect of the Cobalt-N Coordination on the Cobamide Recognition by the Human Vitamin B12 Binding Proteins Intrinsic Factor, Transcobalamin and Haptocorrin." *European Journal of Biochemistry* 199 (2): 299–303. https://doi.org/10.1111/j.1432-1033.1991.tb16124.x.

Subramanian, Veedamali S, Abhisek Ghosal, Rubina Kapadia, Svetlana M Nabokina, and Hamid M Said. 2015. "Molecular Mechanisms Mediating the Adaptive Regulation of Intestinal Riboflavin Uptake Process." *PLoS ONE* 10 (6): 1–13. https://doi.org/10.1371/journal.pone.0131698.

Swigoňová, Zuzana, Al-Walid Mohsen, and Jerry Vockley. 2009. "Acyl-CoA Dehydrogenases: Dynamic History of Protein Family Evolution." *Journal of Molecular Evolution* 69 (2): 176–93. https://doi.org/10.1007/s00239-009-9263-0.

Sybesma, Wilbert, Catherine Burgess, Marjo Starrenburg, Douwe van Sinderen, and Jeroen Hugenholtz. 2004. "Multivitamin Production in Lactococcus Lactis Using Metabolic Engineering." *Metabolic Engineering* 6 (2): 109–15. https://doi.org/10.1016/j.ymben.2003.11.002.

Sybesma, Wilbert F H, 2003. "Metabolic Engineering of Folate Production in Lactic Acid Bacteria." *Thesis Wageningen University*. Wageningen. https://doi.org/10.1007/978-1-4615-0945-5_105.

Sybesma, Wilbert, Marjo Starrenburg, Michiel Kleerebezem, Igor Mierau, Willem M de Vos, and Jeroen Hugenholtz. 2003. "Increased Production of Folate by Metabolic Engineering of Lactococcus Lactis." *Applied and Environmental Microbiology* 69 (6): 3069–76. https://doi.org/10.1128/aem.69.6.3069-3076.2003.

Takenaka, S, S Sugiyama, F Watanabe, K Abe, Y Tamura, and Y Nakano. 1997. "Effects of Carnosine and Anserine on the Destruction of Vitamin B12 with Vitamin C in the Presence of Copper." *Bioscience, Biotechnology, and Biochemistry* 61 (12): 2137–39. https://doi.org/10.1271/bbb.61.2137.

Tamene, Aynadis, Kaleab Baye, Susanna Kariluoto, Minnamari Edelmann, Fabrice Bationo, Nicolas Leconte, and Christèle Humblot. 2019. "Lactobacillus Plantarum P2R3FA Isolated from Traditional Cereal-Based Fermented Food Increase Folate Status in Deficient Rats." *Nutrients* 11 (11). https://doi.org/10.3390/nu11112819.

Taranto, María P, José L Vera, Jeroen Hugenholtz, Graciela F De Valdez, and Fernando Sesma. 2003. "Lactobacillus Reuteri CRL1098 Produces

Cobalamin." *Journal of Bacteriology* 185 (18): 5643–47. https://doi.org/10.1128/jb.185.18.5643-5647.2003.

Taub, Abraham, and Herbert Lieberman. 1953. "Stability of Vitamin B12—Folic Acid Parenteral Solutions." *Journal of the American Pharmaceutical Association (Scientific Ed.)* 42 (4): 183–86. https://doi.org/https://doi.org/10.1002/jps.3030420402.

Thakur, Kiran, Sudhir Kumar Tomar, and Sachinandan De. 2016. "Lactic Acid Bacteria as a Cell Factory for Riboflavin Production." *Microbial Biotechnology* 9(4): 441–51. https://doi.org/10.1111/1751-7915.12335.

Thompson, H O, G Önning, K Holmgren, H S Strandler, and M Hultberg. 2020. "Fermentation of Cauliflower and White Beans with Lactobacillus Plantarum – Impact on Levels of Riboflavin, Folate, Vitamin B12, and Amino Acid Composition." *Plant Foods for Human Nutrition* 75 (2): 236–42. https://doi.org/10.1007/s11130-020-00806-2.

Titcomb, Tyler J, and Sherry A Tanumihardjo. 2019. "Global Concerns with B Vitamin Statuses: Biofortification, Fortification, Hidden Hunger, Interactions, and Toxicity." *Comprehensive Reviews in Food Science and Food Safety* 18 (6): 1968–84. https://doi.org/10.1111/1541-4337.12491.

Torres, Andrea Carolina, Verónica Vannini, Julieta Bonacina, Graciela Font, Lucila Saavedra, and María Pía Taranto. 2016. "Cobalamin Production by Lactobacillus Coryniformis: Biochemical Identification of the Synthetized Corrinoid and Genomic Analysis of the Biosynthetic Cluster." *BMC Microbiology* 16 (1): 240. https://doi.org/10.1186/s12866-016-0854-9.

Trausch, Jeremiah J, Pablo Ceres, Francis E Reyes, and Robert T Batey. 2011. "The Structure of a Tetrahydrofolate-Sensing Riboswitch Reveals Two Ligand Binding Sites in a Single Aptamer." *Structure (London, England : 1993)* 19 (10): 1413–23. https://doi.org/10.1016/j.str.2011.06.019.

Tu, B P, S C Ho-Schleyer, K J Travers, and J S Weissman. 2000. "Biochemical Basis of Oxidative Protein Folding in the Endoplasmic Reticulum." *Science (New York, N.Y.)* 290 (5496): 1571–74. https://doi.org/10.1126/science.290.5496.1571.

Tucker, B J, and R R Breaker. 2005. "Riboswitches as Versatile Gene Control Elements." *Current Opinion in Structural Biology* 15: 342–48.

Turpin, Williams, Christèle Humblot, and Jean Pierre Guyot. 2011. "Genetic Screening of Functional Properties of Lactic Acid Bacteria in a Fermented Pearl Millet Slurry and in the Metagenome of Fermented Starchy Foods." *Applied and Environmental Microbiology* 77 (24): 8722–34. https://doi.org/10.1128/AEM.05988-11.

U.S. Department of Agriculture. 2016. "USDA National Nutrient Database for Standard ReferenceRelease 28."

Vitreschak, Alexey G, Dmitry A Rodionov, Andrey A Mironov, and Mikhail S Gelfand. 2002a. "Regulation of Riboflavin Biosynthesis and Transport Genes in Bacteria by Transcriptional and Translational Attenuation." *Nucleic Acids Research* 30 (14): 3141–51. https://doi.org/10.1093/nar/gkf433.

Vitreschak, Alexey G, Dmitry A Rodionov, Andrey A Mironov, and Mikhail S Gelfand. 2002b. "Regulation of Riboflavin Biosynthesis and Transport Genes in Bacteria by Transcriptional and Translational Attenuation." *Nucleic Acids Research* 30 (14): 3141–51. https://doi.org/10.1093/nar/gkf433.

Wagner-Juaregg, Theodore. 1954. "Chemistry." In *The Vitamins,* edited by W.H. Sebrell and ROBERT S. Harris, 301–32. Academic Press.

Wang, M, D L Roberts, R Paschke, T M Shea, B S Masters, and J J Kim. 1997. "Three-Dimensional Structure of NADPH-Cytochrome P450 Reductase: Prototype for FMN- and FAD-Containing Enzymes." *Proceedings of the National Academy of Sciences of the United States of America* 94 (16): 8411–16. https://doi.org/10.1073/pnas.94.16.8411.

Warren, Martin J, Evelyne Raux, Heidi L Schubert, and Jorge C Escalante-Semerena. 2002. "The Biosynthesis of Adenosylcobalamin (Vitamin B12)." *Natural Product Reports* 19 (4): 390–412. https://doi.org/10.1039/b108967f.

Watanabe, F, and Y Tanioka. 2014. "*Characterization of Corrinoid Compounds in Edible Shellfish.*" In, 413–20.

Watanabe, Fumio, Katsuo Abe, Tomoyuki Fujita, Mashahiro Goto, Miki Hiemori, and Yoshihisa Nakano. 1998. "Effects of Microwave Heating on the Loss of Vitamin B12 in Foods." *Journal of Agricultural and Food Chemistry* 46 (1): 206–10. https://doi.org/10.1021/jf970670x.

Watanabe, Fumio, and Tomohiro Bito. 2018. "Vitamin B(12) Sources and Microbial Interaction." *Experimental Biology and Medicine (Maywood, N.J.)* 243 (2): 148–58. https://doi.org/10.1177/1535370217746612.

Watanabe, Fumio, Hiromi Katsura, Katsuo Abe, and Yoshihisa Nakano. 2000. "Effect of Light-Induced Riboflavin Degradation on the Loss of Cobalamin in Milk." *Journal of Home Economics of Japan* 51 (3): 231–34. https://doi.org/10.11428/jhej1987.51.231.

Wegkamp, Arno, Wietske Van Oorschot, Willem M De Vos, and Eddy J Smid. 2007. "Characterization of the Role of Para-Aminobenzoic Acid Biosynthesis in Folate Production by Lactococcus Lactis." *Applied and Environmental Microbiology* 73 (8): 2673–81. https://doi.org/10.1128/AEM.02174-06.

WHO/FAO. 2006. *Guidelines on Food Fortification with Nutrients*. Edited by Lindsay Allen, Bruno de Benoist, Omar Dary, and Richard Hurrel. WHO Library Cataloguing-in-Publication Data Guidelines. https://doi.org/10.1242/jeb.02490.

Wolf, R, F Reiff, R Wittmann, and J Butzke. 1983. Verfahren zur Herstellung von Riboflavin. Patent EP 0 020 959 B1, issued 1983.

Wolkers-Rooijackers, Judith C M, Martha F Endika, and Eddy J Smid. 2018. "Enhancing Vitamin B12 in Lupin Tempeh by in Situ Fortification." *LWT* 96: 513–18. https://doi.org/https://doi.org/10.1016/j.lwt.2018.05.062.

Xie, Chong, Rossana Coda, Bhawani Chamlagain, Minnamari Edelmann, Paulina Deptula, Pekka Varmanen, Vieno Piironen, and Kati Katina. 2018. "In Situ Fortification of Vitamin B12 in Wheat Flour and Wheat Bran by Fermentation with Propionibacterium Freudenreichii." *Journal of Cereal Science* 81: 133–39. https://doi.org/https://doi.org/10.1016/j.jcs.2018.05.002.

Xie, Chong, Rossana Coda, Bhawani Chamlagain, Pekka Varmanen, Vieno Piironen, and Kati Katina. 2019. "Co-Fermentation of Propionibacterium Freudenreichii and Lactobacillus Brevis in Wheat

Bran for in Situ Production of Vitamin B12." *Frontiers in Microbiology* 10: 1541. https://doi.org/10.3389/fmicb.2019.01541.

Xin, Zhonghao, Lingling Pu, Weina Gao, Yawen Wang, Jingyu Wei, Tala Shi, Zhanxin Yao, and Changjiang Guo. 2017. "Riboflavin Deficiency Induces a Significant Change in Proteomic Profiles in HepG2 Cells." *Scientific Reports* 7 (April): 1–10. https://doi.org/10.1038/srep45861.

Yépez, Alba, Pasquale Russo, Giuseppe Spano, Iuliia Khomenko, Franco Biasioli, Vittorio Capozzi, and Rosa Aznar. 2019. "In Situ Riboflavin Fortification of Different Kefir-like Cereal-Based Beverages Using Selected Andean LAB Strains." *Food Microbiology* 77 (March 2018): 61–68. https://doi.org/10.1016/j.fm.2018.08.008.

Zago, M, E Scaltriti, B Bonvini, M E Fornasari, G Penna, L Massimiliano, D Carminati, M Rescigno, and G Giraffa. 2017. "Genomic Diversity and Immunomodulatory Activity of Lactobacillus Plantarum Isolated from Dairy Products." *Beneficial Microbes* 8 (4): 597–604. https://doi.org/10.3920/BM2016.0223.

Zayas, Carmen L, and Jorge C Escalante-Semerena. 2007. "Reassessment of the Late Steps of Coenzyme B12 Synthesis in Salmonella Enterica: Evidence That Dephosphorylation of Adenosylcobalamin-5'-Phosphate by the CobC Phosphatase Is the Last Step of the Pathway." *Journal of Bacteriology* 189 (6): 2210–18. https://doi.org/10.1128/JB.01665-06.

Zempleni, J, J W Suttie, J F Gregory III, and P J Stover. 2013. *Handbook of Vitamins*. 5th Editio. Boca Raton: CRC Press. https://doi.org/10.1201/b15413.

Zhu, Yun Yang, Kiran Thakur, Jing Yu Feng, Jia Shen Cai, Jian Guo Zhang, Fei Hu, Pasquale Russo, Giuseppe Spano, and Zhao Jun Wei. 2020. "Riboflavin-Overproducing Lactobacilli for the Enrichment of Fermented Soymilk: Insights into Improved Nutritional and Functional Attributes." *Applied Microbiology and Biotechnology* 104 (13): 5759–72. https://doi.org/10.1007/s00253-020-10649-1.

In: Fermented Foods
Editor: Oliver Kovalyov
ISBN: 978-1-53618-656-7

Chapter 2

SULFORAPHANE AS LACTIC ACID BACTERIA - MEDIATED HYDROLYSIS PRODUCT OF GLUCORAPHANIN AND ITS ROLE AS THERAPEUTIC AGENT

Maria K. Syrokou, Markella Tsigkrimani, Spiros Paramithiotis*, Panagiotis N. Skandamis and Eleftherios H. Drosinos

Laboratory of Food Quality Control and Hygiene,
Department of Food Science and Human Nutrition,
Agricultural University of Athens, Athens, Greece

ABSTRACT

Cruciferous vegetables, broccoli, cabbage, cauliflower and vegetable fermented products as well, such as sauerkraut, represent a significant source of sulforaphane (SFN). SFN is a highly bioactive isothiocyanate, derived from its non-bioactive glucosinolate precursor, glucoraphanin

* Corresponding Author's Email: sdp@aua.gr.

(GRP). Upon tissue disruption, β-thioglucosidase enzyme myrosinase or thioglucosidases in intestinal microbiota, convert GRP into SFN and other hydrolysis products. Lactic acid fermentation has been reported to enhance GRP extraction and further SFN bioavailability, through mechanisms yet to be discovered. In addition, several lactic acid bacteria have been reported to possess myrosinase activity, enhancing thus SFN production. SFN metabolism in human and animal cells is performed via mercapturic acid pathway, where SFN is conjugated with glutathione and is subsequently metabolized mainly to its N-acetylcysteine derivative, found in urine and plasma. Its electrophilic nature allows interaction via several signaling cascades, namely Keap1-Nrf2-ARE, MAPK/NF-κB, NLRP3 inflammasome, AMPK, Akt, JAK/STAT, Aβ and tau accumulation, Wnt/β-catenin, Notch and Criptoin prevention or therapy of neurodegenerative, cardiovascular, inflammatory and cancer related diseases. The aim of the present chapter is to elucidate the impact of lactic acid fermentation on SFN rich products and improve our understanding on the role of SFN as potential medicine, in the treatment of various disorders.

INTRODUCTION

Numerous studies have previously reported the inverse correlation between cruciferous vegetable consumption, including broccoli, cabbage and cauliflower with incidence of cancer, neurodegenerative, cardiovascular, inflammatory and many other chronic diseases (Huang et al., 2019; Livingstone et al., 2019). The positive impact of cruciferous vegetable intake on human health, has been attributed to their high content of glucosinolates (GSLs), though themselves are biologically inactive. Upon tissue disruption, during chewing or crushing, GSLs come in contact with coexisting thioglucosidases enzymes myrosinases, usually located in separated plant compartment and are further degraded into hydrolysis products, among which isothiocyanates (ITCs), thiocyanates, nitriles (NITs) and epithionitriles are some of the most commonly found (Luang-In et al., 2018). Sulforaphane (SFN) an ITC derivative of glucoraphanin (GRP), has been extensively studied, due to its health promoting attributes (Silva-Palacios et al., 2019; Wei et al., 2020). Although myrosinases are heat sensitive, thus thermally processed vegetables denaturate these plant enzymes, the presence of myrosinase-like activity in certain human gut bacteria, results in cleavage of GSLs to more bioactive compounds, such as SFN (Narbad & Rossiter, 2018). The protective action of SFN against oxidative stress related diseases, has shifted scientific interest towards search of alternative processes for improvement of SFN formation and stability in food matrices.

Lactic acid fermentation has been carried out for centuries, for preservation and sensory enhancement purposes (Paramithiotis et al., 2017). Over the years, gradual preference for vegetable fermentation by lactic acid bacteria (LAB), compared to more processed foods, results among others, from people's awareness of health promoting potential, stemming from more green approaches. Fermentation of vegetables usually relies on autochthonous LAB species, such as *Lactiplantibacillus plantarum* (formerly *Lactobacillus plantarum*) *and Leuconostoc mesenteroides*, which have been identified as dominant species at different stages of fermentation process, thus exhibit better adaptation to the intrinsic properties of the raw material. Given that LAB constitute the microbiota that drives vegetable fermentation, their metabolic activity coupled with the enzymatic capacity of raw vegetables, can enhance the biological activities of the starting food material (Marco et al., 2017). In other words, the extent to which LAB disrupt cell wall structure and extract bound GSLs, which are further degraded to more bioactive compounds, determines the profile of the final product. Filannino et al., (2015) previously investigating the biotransformation of phenolic acids and flavonoids during LAB fermentation in broccoli puree and cherry juice, reported that LAB exhibited strain-specific metabolism of phenolic acids. However, the feasibility of LAB fermentation on GRP extraction and concomitant SFN production in the food matrix, has been reported by only a few studies, with mechanisms of GRP degradation yet to be discovered (Cai et al., 2019; 2020).

A growing body of evidence highlights the use of phytochemicals, such as SFN, as alternative therapeutic agents against diverse health-related diseases (Pellegrini et al., 2019). Decreased levels of toxicity and side effects coupled with increased efficiency constitute SFN a very promising target for prevention or treatment of carcinogenic, neurodegenerative, inflammatory, respiratory, metabolic and various other types of diseases via modulation of different signaling pathways (Russo et al., 2018; Panieri et al., 2020). Nrf2 activation represents one of the most crucial targets for therapy of oxidative stress related diseases. The electrophilic properties of SFN allow its interaction with many nucleophiles, particularly cysteine residues of Keap1, thus leading to the activation of transcription factor Nrf2 (Uddin et al., 2020). Besides Nrf2 activation, SFN has been shown to modulate MAPK/NF-κB, NLRP3 inflammasome, AMPK, Akt, JAK/STAT, Aβ and tau accumulation, Wnt/β-catenin, Notch and Cripto signaling pathways so as to stimulate diverse protective responses against health-related diseases.

The aim of the present chapter is to investigate the effect of LAB fermentation on SFN formation and stability in vegetable based products and elucidate the mechanisms involved in the protective role of SFN against diverse health related diseases.

SFN Metabolism and Stability

In humans, SFN metabolism in liver and enterocyte starts with conjugation with glutathione, by glutathione *S*-transferase and subsequent degradation predominantly to its N-acetylcysteine derivative, takes place via mercapturic acid pathway, or alternatively is released from glutathione to free SFN (Fahey et al., 2012; Livingstone et al., 2019).

Bioavailability of SFN is mainly determined by the extent to which GSL is degraded to more bioactive compounds during chewing or crashing, prior to passage in liver and gut, where GSL degrading gut bacteria exist. Many factors affect the yield of SFN, the presence or absence of modifier proteins, enzymes, reaction conditions such as pH, temperature and ferrous ions (Angelino et al., 2014; Cai et al., 2020). More accurately, the presence of modifier proteins, such as epithiospecifier protein (ESP) has been reported to drive hydrolysis of GRP towards nitrile formation, a non bioactive compound, however thermal processing of broccoli florets at 60 °C for 5 min, selectively suppressed ESP activity, by shifting the transformation of GRP to SFN production (Latte et al., 2011). In contrast to plants, the presence of ESP has not been reported in bacterial strains. In addition, acidic conditions such as that of stomach, low temperature and ferrous ions, have been shown to promote nitrile formation as well, whereas for pH values ranging between 5 and 8, production of ITCs is typically favoured (Quirante-Moya et al., 2020). Another factor influencing the nature of hydrolysis products is plant myrosinase, the absence of which, shifts the fate of SFN absorption and metabolism towards the efficiency with which human gut microflora hydrolyzes GSLs to more or less bioactive compounds (Narbad & Rossiter, 2018). Finally, the gradual adaptation to *Brassica* vegetables intake, seems to affect the bioavailability of bioactive constituents (Charron et al., 2018).

SFN Formation during LAB Fermentation

Over the past years, the search for myrosinase positive bacteria and their concomitant application in vegetable based fermentation has been gaining much scientific interest. Luang-In et al., (2018) investigated the feasibilty of two GSL degrading bacterial strains belonging to the *Enterococcus casseliflavus* and *Enterobacter xiangfangensis* species, isolated from fermented cabbage and fish, respectively, for further fermentation of Thai cabbage. Compared to spontaneously fermented or unfermented cabbage, induced fermentation led to faster formation of GSL breakdown products, such as SFN, which was attributed to the presence of myrosinase positive bacteria. However, the unstable composition of the natural matrix or a potent antagonism between ITCs and myrosinase positive bacteria negatively influenced ITC stability, over 3 days.

Although many LAB have been reported to carry GSL degrading capacity, namely *Ligilactobacillus agilis* (formerly *Lactobacillus agilis*), *Lb. plantarum*, *Lactobacillus acidophilus* and *Lacticaseibacillus casei*, diverse factors, such as the unstable nature of the food matrix, cooking practices, presence of proper enzymes and acidifying conditions may shift

ITC conversion towards SFN nitrile or nitrile production, rather than SFN (Mullaney et al., 2013; Ye et al., 2019). Cai et al., (2020) reported that application of autochthonous LAB fermentation coupled with mild heat treatment, increased both SFN formation and stability in broccoli puree. Pre-heating procedure at 65 °C/ 3 min, accompanied by LAB fermentation selectively inactivated ESP, which is considered less thermotolerant than myrosinase, while allowed a more gradual acidification in the food matrix, driving the GRP hydrolysis towards SFN formation.

Despite the complexity of the food matrix, SFN formation has been previously documented in fermented cabbage foods, such as sauerkraut (Peñas et al., 2015) and in decreased concentration in kimchi (Kim et al., 1999). Regarding the SFN production from sauerkrauts, two fermentation processes were followed, the former based on autochthonous LAB fermentation and the latter was induced by *Leuconostoc mesenteroides*. As far as the stability of SFN was concerned, both in naturally fermented and *Ln. mesenteroides* sauerkraut, SFN content decreased at the end of the storage period (3 months) so in conventional conditions as in nitrogen (N2)- enriched atmospheres.

SFN-Mediated Signaling Pathways against Health Related Diseases

Over the past few years, the role of SFN as an alternative form of prevention or treatment of carcinogenic, neurodegenerative, inflammatory, metabolic, respiratory, skin, hepatic-related and many other types of diseases, has been receiving more and more attention. Numerous studies have previously shown that SFN- mediated protection involves the activation of the transcription factor Nrf2, which further regulates antioxidant and detoxification responses (Uddin et al., 2020). Although Nrf2 activation is perceived as the core mechanism, via which SFN exerts its health promoting effects, various other signaling cascades are also involved in the regulation of protective responses. Modulation of Keap1-

Nrf2-ARE, MAPK/NF-κB, NLRP3 inflammasome, AMPK, Akt, JAK/STAT, Aβ & tau accumulation, Wnt/β-catenin, Notch and Cripto signaling pathways is considered a key process for prevention or therapy of diverse health related diseases. Table 1 summarizes some recent *in vivo* studies presenting the SFN mediated signaling pathways, via which diverse preventive or therapeutic responses are activated against various health related diseases.

Keap1-Nrf2-ARE Pathway

Nrf2 is a transcription factor, which regulates gene expression under both oxidative and non oxidative conditions, thus activating Nrf2 signaling cascade may confer protective effects towards a great variety of diseases (Matzinger et al., 2018). Nrf2 activation necessitates the presence not only of reactive oxygen species, determining the conditions of oxidative stress, but also that of chemical inducers or co-factors, facilitating electron exchange. In the absence of oxidative stress, Nrf2 is retained in the cytosol, bound to Kelch-like ECH-associated protein 1 (Keap1), subjected to rapid degradation according to a Keap 1 independent ubiquitination model (Zhang et al., 2013). It has been proposed the phosphorylation of Nrf2, at Ser 342 and Ser 347 residues, by glycogen synthase kinase-3 (GSK3β), and the concomitant degradation of Nrf2, through a Cullin1/Rbx1 ubiquitination complex, catalyzed by a β-TrCP, an E3 ubiquitin ligase.

Sulforaphane represents a natural co-factor, whose electrophilic features allow its conjugation with glutathione, by glutathione S-transferases (GSTs), a crucial step for detoxification of electrophiles (Itoh 2010). Upon exposure to electrophile/ oxidative stress, electrophilic SFN enters into the cell and binds to reactive cysteine residues of Keap 1 (Cys151, Cys273, and Cys288), through covalent modification (Kaboli et al., 2020). Conformational changes, lead to the inactivation of Keap1. Liberation of Nrf2 from Keap1 repression, drives the nucleus transfer of Nrf2 and its heterodimerization with small musculoaponeurotic fibrosarcoma (Maf) proteins, MafG, MafK, MafF. Concomitant binding to the antioxidant response elements (AREs) takes place in order to induce the expression of antioxidant and detoxifying enzymes, such as heme oxygenase 1 (HMOX-1), NAD(P)H quinone dehydrogenase 1 (NQO1), heme oxygenase 1 (HO-1), and γ-glutamylcysteine ligase (γ GCL), among many others (Matzinger et al., 2018).

MAPK/ NF-κB Pathways

Mitogen-activated protein kinases (MAPKs) are separated into extracellular signal-regulated kinases 1 and 2, ERK1/2 (P44/P42) and c-Jun N-terminal kinase JNK (P46/P54) and p38 kinase (Wang et al., 2015). Upon oxidative stress conditions, MAPKs stimulate pro-inflammatory transcription factors such as NO, COX-2, IL-1β and TNF-α, which mediate NF-κB signaling cascade (Saedisomeolia et al., 2019). Both MAPK and NF-κB signaling pathways play a vital role in neuronal damage, thus their suppression represents a therapeutic approach. Jin et al., (2019) reported that SFN reduced neuronal damage in LPS mediated BV-2 microglia, by suppressing inhibition of MAPK and NF-κB signaling cascade, via inhibiting phosphorylation of MAPKs, p38, JNK and ERK1/2 and NF-κB p65 as well. A SFN-mediated decrease of protein levels of RIPK3 and MLKL, responsible for necroptosis induction, was also documented, due to MAPK and NF-κB cascade inactivation. Kim et al., (2014) reported that SFN suppressed TPA- mediated NF-κB up-regulation and COX-2 generation in MCF-10A cells via inhibition of ERK1/2-IKKa and NAK–IKKb signaling cascades. In normal conditions, NF-κB is bound to inhibitory kappa Bα protein (IκBs) in the cytosol. When stimulated by pro-inflammatory signals, IκBs are subjected to phosphorylation and concomitant proteasomal degradation, as a result of IκB kinase-dependent, IKK activity (Wang et al., 2009). IKKα and IKKβ constitute two catalytic subunits of IKK complex, which catalyze the phosphorylation of p65 and both p65 and IκB, respectively. SFN treatment has been reported to inhibit the degradation of IκB, with concomitant suppression of NF-κB nuclear translocation and decreased induction of pro-inflammatory genes (Wei et al., 2020).

NLRP3 Inflammasome Pathway

Nucleotide-binding oligomerization domain leucine-rich repeat and pyrin domain-containing protein 3 (NLRP3) inflammasome, has a key role in a series of pathophysiological processes (Bertinaria et al., 2019; Pellegrini et al., 2019). NLRP3 inflammasome is formed by a sensor protein (NLRP3), an adaptor molecule (apoptosis-associated speck-like protein-ASC); and an effector enzyme (caspase-1) (Olcum et al., 2020). Canonical activation of NLRP3 inflammasome normally involves two different steps, the priming and activation step. In the first step, transcription of NLRP3 and pro-IL-1β is upregulated via NF-κB phosphorylation, which has been previously activated via cytokine receptors, such as the tumor necrosis factor (TNFs), Toll-like receptors (TLRs), IL-1 receptor 1 (IL-1R1) and NOD-like receptors (NLRs) (Bertinaria et al., 2019). Signals, such as K^+ efflux, lysosomal, drisruption, ROS, increased levels of cytosolic calcium lead to activation of NLRP3 inflammasome. In the second step, activation of NLRP3 inflammasome leads to the assembly of NLRP3-ASC-pro-caspase-1, which stimulates caspase-1 autoactivation (Yang et al., 2018a). Active caspase-1 cleaves pro-IL-1b and pro-IL-18 to its mature and active form, IL-1b and IL-18, which effectively interfere with the pathophysiology of neurodegenerative disorders (An et al., 2016). The non-canonical pathway of NLRP3 inflammasome activation is mainly regulated by lipopolysaccharide (LPS), a component in many Gram-negative bacteria, generally sensed through Toll-like receptor 4 (TLR4), resulting in nuclear translocation of nuclear NF-κB. Yi et al., (2018), reported the stimulation of caspase-11 as immune response to LPS, via a TLR4- independent manner. Caspase-11 conjugates with lipid A moiety of LPS, thus mediating pyroptosis, an inflammasome type of programmed cell death. Direct inhibition of NLRP3 inflammasome by SFN, has been previously reported (Yang et al., 2018a).

Table 1. Recent *in vivo* studies reporting protective effects of Sulforaphane towards diverse health-related diseases, via modulation of associated signaling pathways

Disease Type	Pathology	Experimental Animal model	Mechanism of action	References
Chemical-induced toxicity	Chromium-induced lung injury	Rats	Activation of Akt/GSK-3β/Fyn p Increased expression of Akt, Nrf2 and its downstream enzymes HO-1 and NQO1 Suppressed activity of GSK-3β & Fyn	Lv et al., 2020
	Arsenic-induced nephrotoxicity	Rats	Activation of PI3K/Akt/Nrf2 p Increased levels of PI3K, Akt and Nrf2 Increased enzyme levels of antioxidants SOD, CAT, GPx, GST	Thangapandiyan et al., 2019
Hepatic	D-galactose-induced liver aging	Rats	Activation of Keap1/Nrf2/HO-1 p Suppressed levels of Keap1 Increased levels of Nrf2 and its downstream antioxidant enzymes CAT, GST & HO-1 Suppressed levels of TNF-α and TGF-β	Saleh et al., 2019
Skin	Atopic dermatitis	Mice	Activation of Nrf2/HO-1 p Suppressed phosphorylation of JAK1/ STAT3 p Increased protein levels of Nrf2, p-Nrf2 and its downstream antioxidant enzyme HO-1 Decreased protein levels of IL-6, IL-1β and TNF-α	Wu et al., 2019
Cardiovascular & Metabolic	Type 2 diabetes-induced cardiomyopathy	Mice	Activation of AMPK/AKT/GSK3β p Enhancement of AMPK mediated lipid metabolism Increased protein levels of Nrf2 Increased protein and mRNA levels of Nrf2 downstream genes CAT & HO-1 Increased expression of PGC-1α, PPAR-α & CPT-1B, attributed to SFN 6M Decreased expression of SCD-1, attributed to SFN 6M	Sun et al., 2020

Disease Type	Pathology	Experimental Animal model	Mechanism of action	References
	Obesity-induced impairment in the male reproductive system	Mice	Inhibition of oxidative stress via up-regulation of Nrf2/ARE p Decreased protein levels of Keap1 Increased protein levels of Nrf2 Recovery of normal autophagy Decreased protein levels of LC3-II/LC3-I and Beclin1	Huo etal., 2019
	Diabetic retinopathy	Rats	Activation of Nrf2 p Suppression of NLRP3 p Decreased levels of TNF-α, IL-6, and IL-1β Increased protein levels of Nrf2 Increased protein and mRNA levels of Nrf2 downstream enzymes HO-1 and NQO1 Decreased expression of NLRP3 inflammasome, cleaved caspase-1 p20, IL-1β p17, and ASC.	Li et al., 2019
	Type 2 diabetes-induced cardiomyopathy	Mice	Activation of LKB1/AMPK p Increased expression of Nrf2 and transcription levels of HO-1 and NQO1 Increased expression of ACC, Sirt1, PGC1-α, ULK1 and LC3 II	Zhang et al., 2014
	Acute Gout Arthritis	Mice	Direct inhibition of NLRP3 p Inhibited caspase-1 activity Reduced level of IL-1β	Yang et al., 2018a
	5-fluorouracil-induced intestinal injury	Mice	Increased mRNA level of *Nrf2*, *HO-1*, *Claudin-1* Decreased mRNA level of *NF-κB* and *iNOS* Suppressed TNF-α level	Wei et al., 2020
Gynecological	Sciatic Nerve Endometriosis	Rats	Activation of Keap1-Nrf2 p Increased levels of Keap1 and Nrf2 Suppressed transcription and protein levels of IL6, IL-1β, TNF-α, COX2 and iNOS	Liu et al., 2020

Table 1. (Continued)

Disease Type	Pathology	Experimental Animal model	Mechanism of action	References
	Endometriosis	Rats	Inhibition of PI3K/Akt p Reduced protein levels of PI3K and Akt Reduced levels of IL-6, IL-10, TNF-α, IFN-g, and VEG Reduced levels of Bcl-2, Bax and cleaved caspase-3 F	Zhou et al., 2019
Respiratory	Acute lung injury	Mice	Activation of Nrf2 dependent p Interaction between Nrf2 and mitochondria Increased expression of genes involved in mitochondrial respiration, energy metabolism and cardio-protection Potential AREs in their DNA sequence	Cho et al., 2019
	Refractory asthma	Mice	Activation of Nrf2 p Increased transcription levels of Nrf2 and its downstream targets, NADPH, NQO1, GCLC, GCLM, HO-1 and GAPDH Elevated expression and activity of HDAC2	Sakurai et al., 2018
Neurodegenerative	Rotenone induced neurotoxicity	Mice	Activation of Nrf2 p Elevated expression of Nrf2 and its downstream targets HO-1 and NQO1 Reversed suppression of mTOR p and neuronal apoptosis Restored suppression of mTOR, p70S6K and 4E-BP1 activity	Zhou et al., 2016
	Scopolamine-induced Amnesia	Mice	Activation of Nrf2/HO-1p Suppressed NF-κB p Increased expression of Nrf2 and its downstream target OH-1 Decreased expression of iNOS, COX-2, TNF-α, IL-6, IL-1β, PGE2, I-κB and pI-κB	Subedi et al., 2019
	Alzheimer disease	Mice	Suppressed protein levels of monomeric and polymeric Aβ, tau and p-tau Higher CHIP and HSP70 levels	Lee et al., 2018

Disease Type	Pathology	Experimental Animal model	Mechanism of action	References
	Alzheimer disease	Mice	Suppressed Aβ oligomer production Lower expression of BACE1 and PS1	Hou et al., 2018
	Alzheimer disease	Mice	Lower level of Aβ deposition Decreased transcription and protein levels of HDAC 1, 2 and 3 Increased transcription and protein levels of p75 neurotrophin receptor Higher acetylation levels of histones H3 and H4 at lysine K9 and K12, respectively	Zhang et al., 2017
	Diffuse axonal injury	Rats	Activation of Nrf2/HO-1 p Increased protein level of Nrf2 and its downstream target HO-1	Wu et al., 2020
	Chronic brain ischemic injury	Rats	Activation of Nrf2/HO-1 p Increased level of Nrf2 and its downstream target HO-1	Mao et al., 2018
	Cerebral Ischemic/Reperfusion Injury	Rats	Inhibition of NLRP3 inflammasome p Suppressed transcription level of NLRP3 Decreased expression of cleaved caspase-1, IL-1β and IL-18	Yu et al., 2017
	3-Nitropropionic Acid-Induced Striatal Toxicity	Mice	Activation of Keap1/Nrf2/ARE p Suppression of the MAPK and NF-κB p Decreased transcription level of Keap1 Increased transcription and protein levels of Nrf2 and HO-1 Increased transcription levels of NQO-1, GCLs Decreased protein levels of p-p38, p65, p50, p-IκBα Decreased transcription levels of TNF-α, IL-1β and IL-6 Decreased transcription and protein levels of iNOS, and COX-2	Jang and Cho, 2016
Cancer	Breast cancer	Mice	Targeting Cripto-mediated p Inhibited interaction of CR1 with both Alk4 and GRP78 Decreased transcription levels of CR1, FOXD3, Wnt3 and Notch4	Castro et al., 2019

Table 1. (Continued)

Disease Type	Pathology	Experimental Animal model	Mechanism of action	References
	Lung cancer EMT	Mice	Activation of ERK5 Increased protein levels of ERK5 and p-ERK5 Decreased protein levels of N-cadherin and Snail1	Chen et al., 2019
	Skin cancer	Mice	Inhibition of sulfatase-2 Increased transcription level of Nrf2 Decreased transcription levels of NF-κB and caspase-3 Decreased transcription and protein levels of TNF-α and IL-1β	Alyoussef and Taha, 2018
	Prostate cancer	Mice	Inhibition of HDAC3 protein expression Decreased *p16* transcription, H3K9 me3 and H3K18ac levels	Beaver et al., 2018

p.: pathway; SFN: sulforaphane;Nrf2: nuclear factor erythroid 2 related factor 2; Akt: protein kinase B (PKB); GSK-3β: glycogen synthase kinase 3 beta; Fyn: tyrosine protein kinase; HO-1: heme oxygenase-1; NQO-1: NAD(P)Hquinone oxidoreductase-1; PI3K: phosphatidylinositol 3-kinase; SOD: superoxide dismutase; CAT: catalase; GPX: glutathione peroxidase; GST: glutathione S-transferase; Keap1: Kelch-like ECH-associated protein 1; TNF-*α*: tumor necrosis factor-alpha; TGF-β: transforming growth factor-beta; JAK1: Janus kinase 1; STAT: Signal Transducers and Activators of Transcription; IL-6: interleukin 6; IL-1β: interleukin 1 beta; AMPK: AMP-activated protein kinase; PGC-1*α*: PPAR-gamma coactivator-1 alpha; PPAR-*α* :peroxisome proliferator activated receptor alpha; CPT-1B: carnitine palmitoyltransferase-1B; SCD-1: stearyl coenzyme A decarboxylase; LC3: light chain 3; NLRP3: Nucleotide-binding oligomerization domain leucine-rich repeat and pyrin domain-containing protein 3; ASC: apoptosis associated speck-like protein; LKB1: Liver kinase B1; ACC: acetyl-CoA carboxylase; Sirt1: sirtuin 1; ULK1: Unc-51 like autophagy activating kinase 1; NF-κB: Nuclear factor κB; iNOS: inducible NO synthase; IL-10: interleukin 10; IFN: interferon; VEGF: vascular endothelial growth factor; Bcl-2: B-cell lymphoma/leukemia 2; AREs: antioxidant response elements; GCLC: glutamate-cysteine ligase catalytic subunit; GCLM: glutamate-cysteine ligase modifier subunit; GAPDH: glyceraldehyde 3-phosphate dehydrogenase; HDAC2: histone deacetylase 2; mTOR: mammalian target of rapamycin; p70S6K: 70 S6 kinase; 4E-BP1: eukaryotic initiation factor 4E (eIF4E) binding protein 1; PGE2: prostaglandin E2; Aβ: amyloid-β; HSP: heat shock protein; CHIP: C-terminus of HSP70-interacting protein; BACE1: Beta-secretase 1; PS1: presenilin1; IL-18: interleukin 18; MAPK: mitogen-activated protein kinase; p-p38: phosphorylated p-38; p-IκBα: phosphorylated IκBα; p65, p50, p-IκBα, IκBα: proteins of Nf-κB; CR1: CRIPTO-1/TDGF1 gene; AlLK-4: Activin receptor type-1B (ACVR1B); GRP78: 78-kDa glucose-regulated protein; FOXD3: Forkhead box D3; ERK5: extracellular signal regulated kinase 5; Snail1: zinc finger protein, H3K9me3; trimethyl-histone H3 lysine 9; H3K18ac, acetyl-histone H3 lysine 18.

AMPK Pathway

Adenosine monophosphate-activated protein kinase (AMPK) pathway represents a potential therapeutic strategy for neurodegenerative diseases related to oxidative stress (Lee et al., 2014). Zhang et al., (2014) reported that in Type 2 diabetes mellitus mice, SFN activated Liver kinase B1 (LKB1), a major upstream kinase, which up-regulated AMPK, via its activating loop at Thr-172 of the α subunit. Activated AMPK phosphorylates Unc-51-like kinase 1 (ULK1) or controls ULK1 via suppression of mTOR complex 1, (mTORC1) (Shah et al., 2017). Thus activated AMPK regulates fatty acid oxidation (FAO) in skeletal muscle by activating peroxisome proliferator-activated receptors (PPARs), ligand-inducible transcription factors (Bougarne et al., 2018; Sun et al., 2020). Among PPARs, PPARa, when co-activated with PPARγ coactivator (PGC-1), regulates expression of genes involved in mitochondrial FAO. Up-regulation of PPARs leads to concomitant stimulation of malonyl-CoA decarboxylase (MCD). Malonyl CoA, an allosteric inhibitor of carnitine palmitoyltransferase (CPT-1), catalyzes the transfer of acyl groups of long-chain fatty acyl-CoA into the mitochondria, thus its degradation leads to increased FAO. Finally activated AMPK, followed by increased levels of NAD concentrations, led to activation of sirtuins (SIRTs), NAD^+ dependent deacetylating enzymes (Connell et al., 2019). SIRT1, the most studied SIRT, deacetylated/ activated non histone proteins, such as PGC-1α. PGC-1α is a key regulator of fatty acid oxidation and energy metabolism.

Akt Pathway

PI3K/Akt/mTOR

Mammalian target of rapamycin (mTOR) represents a downstream target of PI3K/Akt cascade, whose induction, regulates cell growth, proliferation, metabolism and survival (Chang et al., 2017). Dysregulation of mTOR is closely related to various diseases, thus its inhibition is considered a potential therapeutic target (Zhang et al., 2019). In the presence of growth factors or cytokines, receptor tyrosine kinases are activated, thereby up-regulating phosphatidylinositol 3′-kinase (PI3K) (Ferrara- Romeo et al., 2020). PI3K converts PIP2 to PIP3, leading to the phosphorylation of protein kinase B (Akt). Lee et al., (2020) reported that activated Akt induces mTORC1 (one of the two complexes of mTOR), thus its downstream targets, via up-regulating ribosomal protein S6 kinase (S6K) and down-regulating eukaryotic initiation factor 4E binding protein-1 (4EBP1). Previous studies have reported that induction of autophagy is mediated via inactivating PI3K/Akt/mTOR signaling cascade (Shao et al., 2020; Chang et al., 2017).

PTEN/PI3K/Akt

Phosphatase and tensin homolog (PTEN) is referred to as silence tumor suppressor with key role in various cancer-related diseases and significant impact on stimulation of autophagy (Kaboli et al., 2020). Activation of PTEN is crucial for inhibition of PI3K/Akt signaling cascade. Yang et al., (2018b) reported that SFN decreased HDAC6 expression, which catalyzes deacetylation of PTEN at Lys 163, thus leading to PTEN membrane translocation and concomitant up-regulation. Activated PTEN inhibited the phosphorylated proteins, pPI3K and pAkt, thus stimulating autophagy in triple negative breast cancer cells.

Akt/GSK3β

AMPK, a sensor of cellular energy fluctuates, when activated, phosphorylates Akt, which in turns suppresses glycogen synthase kinase 3 beta (GSK3β) by phosphorylating the Ser9 residue, thus promoting nuclear transportation of Nrf2 and concomitant expression of antioxidant and detoxifying enzymes (Sun et al., 2020). Lv et al., (2020) reported the SFN's ameliorating effects on chromium-induced lung injury in rats via activation of Akt/GSK3β/Fyn cascade. Tyrosine kinase Fyn, a negative regulator of Nrf2, when activated by GSK3β, phosphorylates tyrosine 568 of Nrf2, leading to Nrf2 nucleus export and concomitant degradation (Jain

& Jaiswal, 2007). Thus, inhibition of GSK3β/ Fyn cascade, promotes nucleus translocation of Nrf2.

JAK/STAT Pathway

Janus Kinase/Signal Transducer and Activator of Transcription (JAK/STAT) pathway regulates signal transduction from cell membrane into the nucleus and is primarily activated by cytokines or growth factors (Nabavi et al., 2019). Several studies have reported that induction of JAK/STAT signaling cascade has been linked to cancer and neuroinflammatory diseases (Yan et al., 2018; Xin et al., 2020). Thus inhibition of JAK/ STAT pathway of natural compounds, such as curcumin and quercetin, has been proposed as potential therapeutic strategy (Nabavi et al., 2019; Khan et al., 2020). Activation of JAK/ STAT signaling cascade is mediated by attachment of cytokines or growth factors to the receptors of cell membrane, which up-regulates JAKs. Activated JAKs phosphorylate a tyrosine residue in the cytoplasmic domain of cytokine receptor and generates a docking site for phosphorylation of STATs. Activated STATs are dimerized and translocate to the nucleus, where they induce the expression of target genes (Xin et al., 2020). SFN mediated inhibition of JAK1/ STAT3 signaling cascade in atopic dermatitis mice model, was previously demonstrated (Wu et al., 2019).

Aβ and Tau Accumulation

Extracellular aggregation of amyloid beta (Aβ) peptides and intracellular accumulation of hyperphosphorylated tau, with concomitant formation of senile plaques and neurofibrillary tangles, respectively, play a crucial role in pathology of Alzheimer disease (Uddin et al., 2020). Although the precise mechanism of action of SFN in AD prevention needs further investigation, inhibition of Aβ or tau aggregation, has been proposed as potential therapeutic strategies (Huang et al., 2019). Zhang et al., (2017) reported that SFN treatment exerted neuroprotective effects and reduced Aβ deposition in AD mice model. The anti-Aβ mode of action of SFN may involve up-regulation of p75 neurotrophin receptor, via inhibiting histone deacetylase 1 and 3 expressions. In vivo study of Lee et al., (2018) confirmed the ameliorating effects of SFN against Aβ toxicity in AD mice model, via up-regulating the heat shock protein HSP70 and co-chaperone CHIP, thus leading to clearance of Aβ and tau accumulation. Both HSP70 and CHIP are capable of binding tau and stimulating ubiquitin degradation signaling cascade, catalyzed by E3 ligase activity. Hou et al., (2018), reported another possible mode of action of SFN against Aβ and tau aggregation, via indirect inhibition of oxidative stress and inflammation, due to decreased levels of TNF-a, IL-1β in AD transgenic mice. Another explanation for decreased levels of Aβ, was considered the inhibitory effect of SFN on BACE1 and PS1 expression, enzymes responsible for Aβ production.

Canonical Wnt/β-Catenin Pathway

Canonical Wnt/β-catenin signaling cascade is considered a key regulator of cancer stem cell (CSC) generation and maintenance (Li et al., 2019a). Under normal conditions, where Wnt signaling is absent, cytoplasmic β-catenin is bound to a protein complex comprising of kinases GSK3β and casein kinase 1α (CK1α), adenomatous polyposis coli (APC) and AXIN (Zhan et al., 2017). GSK3β phosphorylates β-catenin at amino acids Ser33/Ser37/Thr41, thus promoting its proteasomal degradation (Li et al., 2013). In the presence of Wnt signaling, Wnt proteins interact with Frizzled receptors (FZD) and Lipoprotein receptor-related protein (LRP), resulting in β-catenin translocation into the nucleus (Cianciosi et al., 2018). Nuclear β-catenin interacts with Lymphoid enhancer factor (LEF) and T-cell factor (TCF), leading to the up-regulation of many target genes associated with the self-renewal behavior of CSCs. Li et al. (2010) reported that SFN decreased breast CSCs in vitro and in vivo and inhibited

Wnt/β-catenin pathway, via inducing phosphorylation of GSK3β and thus proteasomal degradation of β-catenin.

Notch Pathway

Notch pathway plays a fundamental role in differentiation and preservation of stem cells, thus its activation has been linked to cancer progression (Dandawate et al., 2016; Nalla et al., 2019). In mammals, Notch cascade is comprised of four transmembrane receptors Notch (1-4) and five ligands (Jagged 1, Jagged 2, Dll1, Dll3, and Dll4). A series of cleavage steps leads to the up-regulation of Notch signaling cascade (Krishna et al., 2019). At first, Notch receptors undergo proteolytic degradation by Furin-like convertases in the Golgi complex and concomitant translocation to the cell surface occurs. Interactions between Notch receptors and ligands result in further hydrolysis by metalloprotease 10 (ADAM10) and TNFα converting enzyme (TACE). γ-secretase catalyzes a last cleavage, resulting in the generation of Notch intracellular domain (NICD) in the cytosol and its subsequent nuclear translocation (Rangel et al., 2016). Nuclear NICD heterodimerizes with CSL, via which interaction, the regulation of target Notch genes is accomplished (Canciosi et al., 2018). Kallifatidis et al., (2011) reported that SFN improved the anti-cancer effects of chemotherapeutic agents against pancreatic and prostate cancer stem cells, via suppressing the expression of Notch-1, which contributed to a further prevention of NF-κB pathway.

Cripto Pathway

Cripto-1, also known as Teratocarcinoma-derived growth factor1 (TDGF-1), plays a key role in the regulation of different types of cancer (Klauzinska et al., 2014). Cripto-1 generally serves during embryogenesis, as a co-factor for the transforming growth factor-β (TGF-β) family of ligands Nodal growth and differentiation factors (GDFs) 1 and 3 and glucose regulated protein, 78 kDa (GRP78), as well, thus inducing the up-regulation of type I receptors Alk4 and Alk7 or a type II receptor complex formation, with concomitant induction of Smad-dependent signaling cascade (Rangel et al., 2016). In addition, Cripto-1 functions as a co-factor in the regulation of Wnt/β catenin pathway, also involved in cell migration, during embryogenesis and different stages of cancer development (Gudbergsson et al., 2019). Nagaoka et al., (2013) documented the cross-talk between Cripto-1 and Wnt/β catenin cascade, through binding of Cripto-1 to lipoprotein receptor-related proteins (LRP) 5 and 6, which further facilitates Wnt3 binding. Castro et al., (2019) reported the ameliorating effects of SFN in Triple- negative breast cancer stem- like cells, via inhibiting the complex formation of Cripto-1 with both Alk4 and GRP78.

Conclusion

The health promoting attributes of sulforaphane, as well as the related mechanisms, have been extensively studied and well characterized. However, the effect of lactic acid fermentation on the concentration, stability and bioavailability of this phytochemical has not been yet adequately assessed. Given its importance in disease prevention and treatment, thorough study of the effect of lactic acid fermentation is expected to take place in the near future.

References

Alyoussef, A. & Taha, M. (2019). Antitumor activity of sulforaphane in mice model of skin cancer via blocking sulfatase-2. *Experimental Dermatology, 28*, 28–34.

An, Y. W., Jhang, K. A., Woo, S. Y., Kang, J. L. & Chong, Y. H. (2016). Sulforaphane exerts its anti-inflammatory effect against amyloid-β peptide via STAT-1 dephosphorylation and activation of Nrf2/HO-1 cascade in human THP-1 macrophages. *Neurobiology of Aging 38*, 1–10.

Angelino, D. & Jefery, E. (2014). Glucosinolate hydrolysis and bioavailability of resulting isothiocyanates: focus on glucoraphanin. *Journal of Functional Foods 7*, 67–76.

Beaver, L. M., Löhr, C. V., Clarke, J. D., Glasser, S. T., Watson, G. W., Wong, C. P, Zhang, Z., Williams, D. E., Dashwood, R. H., Shannon J, Thuillier, P. & Ho, E. (2018). Broccoli sprouts delay prostate cancer formation and decrease prostate cancer severity with a concurrent decrease in HDAC3 protein expression in transgenic adenocarcinoma of the mouse prostate (TRAMP) mice. *Current Development of Nutrition, 2*, 002.

Bertinaria, M., Gastaldi, S., Marini, E. & Giorgis, M. (2019). Development of covalent NLRP3 inflammasome inhibitors: chemistry and biological activity, *Archives of Biochemistry and Biophysics, 670*, 116-139.

Bougarne, N., Weyers, B., Desmet, S. J., Deckers, J., Ray, D. W., Staels, B., De Bosschers, K. (2018) Molecular actions of PPARα in lipid metabolism and inflammation. *Endocrine Reviews, 39*, 760–802.

Cai, Y. X., Augustin, M. A., Jegasothy, H., Wang, J. H. & Terefe, N. S. (2020). Mild heat combined with lactic acid fermentation: a novel approach for enhancing sulforaphane yield in broccoli puree. *Food & Function, 11*, 779–786.

Cai, Y. X., Wang, J. H. McAuley, C., Augustin M. A. & Terefe, N. S. (2019). Fermentation for enhancing the bioconversion of glucoraphanin into sulforaphane and improve the functional attributes of broccoli puree. *Journal of Functional Foods, 61*, 103461.

Castro, N. P., Rangel, M. C., Merchant, A. S., MacKinnon, G. M., Cuttitta, F., Salomon, D. S. & Kim, Y. S. (2019). Sulforaphane suppresses the growth of triple-negative breast cancer stem-like cells in vitro and in vivo. *Cancer Prevention Research, 12*, 147–158.

Chang, H., Li, X., Cai, Q., Li, C., Tian, L., Chen, J., Xing, X., Gan, Y., Ouyang, W. & Yang, Z. (2017). The PI3K/Akt/mTOR pathway is involved in CVB3-induced autophagy of HeLa cells. *International Journal of Molecular Medicine, 40*, 182–192.

Charron, C. S., Vinyard, B. T., Ross, S. A., Seifried, H. E., Jeffery, E. H. & Novotny, J. A. (2018). Absorption and metabolism of isothiocyanates formed from broccoli glucosinolates: effects of BMI and daily consumption in a randomised clinical trial. *British Journal of Nutrition, 120*, 1370–1379.

Chen, Y., Chen, J., Ge, M., Zhang, Q., Wang, X., Zhu, J., Xie, C., Li, X, Zhong, C & Han, H. (2019). Sulforaphane inhibits epithelial–mesenchymal transition by activating extracellular signal-regulated kinase 5 in lung cancer cells. *Journal of Nutritional Biochemistry, 72*, 108219.

Cho, H. Y., Miller-DeGraff, L., Blankenship-Paris, T., Wang, X., Bell, D. A., Lih, F., Deterding, L., Panduri, V., Morgan, D. L., Yamamoto, M., Reddy, A. J., Talalay, P. & Kleeberger, S. R. (2019). Sulforaphane enriched transcriptome of lung mitochondrial energy metabolism and provided pulmonary injury protection via Nrf2 in mice. *Toxicology and Applied Pharmacology, 364*, 29–44.

Cianciosi, D., Varela-Lopez, A., Forbes-Hernandez, T. Y., Gasparrini, M., Afrin, S., Reboredo-Rodriguez, P., Zhang, J., Quiles, J. L, Nabavi, S. F., Battino, M & Giampieri, F. (2018). Targeting molecular pathways in cancer stem cells by natural bioactive compounds. *Pharmacological Research, 135*, 150–165.

Connell, N. J., Houtkooper, R. H. & Schrauwen, P. (2019). NAD(+) metabolism as a target for metabolic health: Have we found the silver bullet? *Diabetologia, 62*, 888–899.

Dandawate, P. R., Subramaniam, D., Jensen, R. A. & Anant, S. (2016). Targeting cancer stem cells and signaling pathways by phytochemicals: Novel approach for breast cancer therapy. *Seminars in Cancer Biology, 40-41*, 192–208.

Fahey, J. W., Wehage, S. L., Holtzclaw, W. D., Kensler, T. W., Egner, P. A., Shapiro, T. A. & Talalay, P. (2012). Protection of humans by plant glucosinolates: Efficiency of conversion of glucosinolates to isothiocyanates by the gastrointestinal microflora. *Cancer Prevention Research, 5*, 603–611.

Ferrara-Romeo, I., Martinez, P., Saraswati, S., Whittemore, K., Graña-Castro, O., Poluha, L. T., Serrano, R., Hernandez-Encinas, E., Blanco-Aparicio, C., Flores, J. M. & Blasco, M. A. (2020). The mTOR pathway is necessary for survival of mice with short telomeres. *Nature Communications, 11*, 1168.

Filannino, P., Bai, Y., Di Cagno, R., Gobbetti, M. & Gänzle, M. G. (2015). Metabolism of phenolic compounds by *Lactobacillus* spp. during fermentation of cherry juice and broccoli puree. *Food Microbiology, 46*, 272–279.

Gudbergsson, J. M. & Duroux, M. (2019). Cripto-1 localizes to dynamic and shed filopodia associated with cellular migration in glioblastoma cells. *European Journal of Cell Biology, 98*, 151044.

Hou, T. T., Yang, H. Y., Wang, W., Wu, Q. Q., Tian, Y. R. & Jia, J. P. (2018). Sulforaphane inhibits the generation of amyloid-β oligomer and promotes spatial learning and memory in Alzheimer's disease (PS1V97L) transgenic mice. *Journal of Alzheimers Disease, 62*, 1803–1813.

Huang, C., Wu, J., Chen, D., Jin, J., Wu, Y. & Chen, Z. (2019). Effects of sulforaphane in the central nervous system. *European Journal of Pharmacology, 853*, 153–168.

Huo, L., Su, Y., Xu, G., Zhai, L. & Zhao, J. (2019). Sulforaphane protects the male reproductive system of mice from obesity-induced damage: Involvement of oxidative stress and autophagy. *International Journal of Environmental Research and Public Health, 16*, 3759.

Itoh, K., Mimura, J. & Yamamoto, M. (2010). Discovery of the negative regulator of Nrf2, Keap1: a historical overview. *Antioxidants and Redox Signalling, 13*, 1665–78.

Jain, A. K. & Jaiswal, A. K. (2007). GSK-3beta acts upstream of Fyn kinase in regulation of nuclear export and degradation of NF-E2 related factor 2. *Journal of Biological Chemistry, 282*, 16502-16510.

Jang, M. & Cho, I. H. (2016). Sulforaphane ameliorates 3-nitropropionic acid-induced striatal toxicity by activating the Keap1-Nrf2-ARE pathway and inhibiting the MAPKs and NF-κB pathways. *Molecular Neurobioly, 53*, 2619–2635.

Jin, X., Liu, M. Y., Zhang, D. F., Zhong, X., Du, K., Qian, P., Gao, H. & Wei, M. J. (2019). Natural products as a potential modulator of microglial polarization in neurodegenerative diseases. *Pharmacological Research, 145*, 28.

Kaboli, P. J., Khoshkbejari, M. A., Mohammadi, M., Abiri A., Mokhtarian, R., Vazifemand, R., Amanollahi, S., Sani, S. Y., Li, M., Zhao, Y., Wu, X., Shen, J., Cho, C. H. & Xiao, Z. (2020). Targets and mechanisms of sulforaphane derivatives obtained from cruciferous plants with special focus on breast cancer - contradictory effects and future perspectives, *Biomedicine & Pharmacotherapy, 121*, 109635.

Kallifatidis, G., Labsch, S., Rausch, V., Mattern, J., Gladkich, J., Moldenhauer, G., Buchler, M. W., Salnikov, A. V. & Herr, I. (2011). Sulforaphane increases drug-mediated cytotoxicity toward cancer stem-like cells of pancreas and prostate. *Molecular Therapy, 19*, 188–195.

Khan, A. Q., Ahmed, E. I., Elareer, N., Fathima, H., Prabhu, K. S., Siveen, K. S., Kulinski, M., Azizi, F., Dermime, S., Ahmad, A., Steinhoff, M. & Uddin, S. (2020). Curcumin-mediated apoptotic cell death in papillary thyroid cancer and cancer stem-like cells through targeting of the JAK/STAT3 signaling pathway. *International Journal of Molecular Sciences, 21*, 438.

Kim, H. Y., Hwang, K. W. & Park, S. Y. (2014). Extracts of Actinidia arguta stems inhibited LPS induced inflammatory responses through nuclear factor-kappaB pathway in Raw 264.7 cells. *Nutrition Research, 11*, 1008–1016.

Kim, J., Lee, S., Choi, B. R., Yang, H., Hwang, Y., Park, J. H. Y., LaFerla, F. M, Han, J. S., Lee, K. W. & Kim, J. (2017) Sulforaphane epigenetically enhances neuronal BDNF expression and TrkB signaling pathways. *Molecular Nutrition and Food Research, 61*, 1600194.

Kim, M. R., Lee, K. J., Kim, H. Y., Kim, J. H., Kim, Y. B. & Sok, D. E. (1999). Effect of various kimchi extracts on the hepatic glutathione S-transferase activity of mice. *Food Research International, 31*, 389–394.

Kim, S., Viswanath, A. N. I., Park, J.-H, Lee, H. E., Park, A. Y., Choi, J. W., Kim, H. J., Londhe, A. M., Jang, B. K., Lee, J., Hwang, H., Lim, S. M., Pae, A. N. & Park, K. D. (2020). Nrf2 activator via interference of Nrf2-Keap1 interaction has antioxidant and anti-inflammatory properties in Parkinson's disease animal model. *Neuropharmacology, 167*, 107989.

Klauzinska, M., Castro, N. P., Rangel, M. C., Spike, B. T., Gray, P. C., Bertolette, D., Cuttitta, F. & Salomon, D. (2014). The multifaceted role of the embryonic gene Cripto-1 in cancer, stem cells and epithelial-mesenchymal transition. *Seminars in Cancer Biology, 29*, 51 – 58.

Krishna, B. M., Jana, S., Singhal, J., Horne, D., Awasthi, S., Salgia, R. & Singhal, S. S. (2019). Notch signaling in breast cancer: from pathway analysis to therapy, *Cancer Letters, 461*, 123–131.

Latte, K. P., Appel K. E. & Lampen, A. (2011). Health benefits and possible risks of broccoli - an overview. *Food and Chemical Toxicology, 49*, 3287–3309.

Lee, J. H., Jeong, J. K. & Park, S. Y. (2014). Sulforaphane-induced autophagy flux prevents prion protein-mediated neurotoxicity through AMPK pathway. *Neuroscience, 278*, 31–39.

Lee, M. J., Jin, N., Grandis, J. R. & Johnson, D. E. (2020). Alterations and molecular targeting of the GSK-3 regulator, PI3K, in head and neck cancer. *BBA - Molecular Cell Research, 1867*, 118679.

Lee, S., Choi, B. R., Kim, J., La Ferla, F. M., Park, J. H. Y., Han, J. S., Lee, K. W. & Kim, J. (2018). Sulforaphane upregulates the heat shock protein co-chaperone CHIP and clears amyloid-β and Tau in a mouse model of Alzheimer's disease. *Molecular Nutrition Food Research, 62*, 1800240.

Li, S., Yang, H. & Chen, X. (2019b). Protective effects of sulforaphane on diabetic retinopathy: Activation of the Nrf2 pathway and inhibition of NLRP3 inflammasome formation. *Experimental Animals, 68*, 221–231.

Li, Y. & Zhang, T. (2013). Targeting cancer stem cells with sulforaphane, a dietary component from broccoli and broccoli sprouts. *Future Oncology, 9*, 1097–1103.

Li, Y., Xiao, Y., Lin, H. P., Reichel, D., Bae, Y., Lee, E. Y., Jiang, Y., Huang, X., Yang & C., Wang, Z. (2019a). In vivo β-catenin attenuation by the integrin α5-targeting nanodelivery strategy suppresses triple negative breast cancer stemness and metastasis. *Biomaterials, 188*, 160–172.

Li, Y., Zhang, T., Korkaya, H., Liu, S., Lee, H.-F., Newman, B., Yu, Y., Clouthier, S. G., Schwartz, S. J. & Wicha, M. S. (2010). Sulforaphane, a dietary component of broccoli/broccoli sprouts, inhibits breast cancer stem cells. *Clinical Cancer Research, 16*, 2580–2590.

Liu, Y., Zhang, Z., Lu, X., Meng, J., Qin, X. & Jiang, J. (2020). Anti-nociceptive and anti-inflammatory effects of sulforaphane on sciatic endometriosis in a rat model. *Neuroscience Letters, 723*, 134858.

Livingstone, T. L., Beasy, G., Mills, R. D., Plumb, J., Needs, P. W., Mithen, R. & Traka, M. H. (2019). Plant bioactives and the prevention of prostate cancer: evidence from human studies. *Nutrients, 11*, 2245.

Luang-In, V., Deeseenthum, S., Udomwong, P., Saengha, W. & Gregori, M. (2018). Formation of sulforaphane and iberin products from Thai cabbage fermented by myrosinase-positive bacteria. *Molecules, 23*, 955.

Lv, Y., Jiang, H., Li, S., Han, B., Liu, Y., Yang, D., Li, Y., Yang, Q., Wu, P & Zhang, Z. (2020). Sulforaphane prevents chromium-induced lung injury in rats via activation of the Akt/GSK-3b/Fyn pathway. *Environmental Pollution, 259*, 113812.

Mao, L., Yang, T., Li, X., Lei, X., Sun, Y., Zhao, Y., Zhang, W., Gao, Y., Sun, B. & Zhang, F. (2018). Protective effects of sulforaphane in experimental vascular cognitive impairment: Contribution of the Nrf2 pathway. *Journal of Cerebral Blood Flow & Metabolism, 39*, 352-366.

Marco, M. L., Heeney, D., Binda, S., Cifelli, C. J., Cotter, P. D., Foligné, B., Gänzle, M., Kort, R., Pasin, C., Pihlanto, A., Smid E. J. & Hutkins, R. (2017). Health benefits of fermented foods: microbiota and beyond. *Current Opinion in Biotechnology, 44*, 94–102.

Matzinger, M., Fischhuber, K. & Heiss, E. H. (2018). Activation of Nrf2 signaling by natural products-can it alleviate diabetes? *Biotechnology Advances, 36*, 1738–1767.

Mullaney, J. A., Kelly, W. J., McGhie, T. K., Ansell, J. & Heyes, J. A. (2013). Lactic acid bacteria convert glucosinolates to nitriles efficiently yet differently from Enterobacteriaceae. *Journal of Agricultural and Food Chemistry, 61*, 3039–3046.

Nabavi, S. M., Ahmed, T., Nawaz, M., Devi, K. P., Balan, D. J., Pittalà, V., Argüelles-Castilla, S., Testai, L., Khan, H., Sureda, A., de Oliveira, M. R., Vacca, R. A., Xu, S., Yousefi, B., Curti, V., Daglia, M., Sobarzo-Sánchez, E., Filosa, R., Nabavi, S. F., Majidinia, M., Dehpour, A. R. & Shirooie, S. (2019). Targeting STATs in neuroinflammation: The road less traveled! *Pharmacological Research, 141*, 73–84.

Nagaoka, T., Karasawa, H., Turbyville, T., Rangel, M.-C., Castro, N. P., Gonzales, M., Baker, A., Seno, M., Lockett, S., Greer, Y. E., Rubin, J. S., Salomon, D. S. & Bianco, C. (2013). Cripto-1 enhances the canonical Wnt/β-catenin signaling pathway by binding to LRP5 and LRP6 co-receptors. *Cellular Signalling, 25*, 178–189.

Nalla, L. V., Kalia, K. & Khairnar, A. (2019). Self-renewal signaling pathways in breast cancer stem cells. *International Journal of Biochemistry and Cell Biology, 107*, 140–153.

Narbad, A. & Rossiter, J. T. (2018). Gut glucosinolate metabolism and isothiocyanate production. *Molecular Nutrition Food Research 62*, 1700991.

Olcum, M., Tastan, B., Kiser, C., Genc, S. & Genc, K. (2020). Microglial NLRP3 inflammasome activation in multiple sclerosis. *Advances in Protein Chemistry and Structural Biology, 119*, 247–308.

Panieri, E., Buha, A., Telkoparan-Akillilar, P., Cevik, D., Kouretas, D., Veskoukis, A., Skaperda, Z., Tsatsakis, A., Wallace, D., Suzen, S. & Saso, L. (2020). Potential applications of NRF2 modulators in cancer therapy. *Antioxidants, 9*, 193.

Paramithiotis, S., Papoutsis, G. & Drosinos, E. H. (2017). Lactic acid fermentation of fruits and vegetables: an overview. In: Paramithiotis S.

(ed) *Lactic acid fermentation of fruits and vegetables.* CRC Science, Boca Raton, pp 1–18.

Pellegrini, C., Fornai, M., Antonioli, L., Blandizzi, C. & Calderone, V. (2019). Phytochemicals as novel therapeutic strategies for NLRP3 inflammasome-related neurologica, metabolic, and inflammatory diseases. *International Journal of Molecular Science, 20*, 2876.

Peñas, E., Martínez-Villaluenga, C., Pihlava, J.-M. & Frias, J. (2015). Evaluation of refrigerated storage in nitrogen-enriched atmospheres on the microbial quality, content of bioactive compounds and antioxidant activity of sauerkrauts. *LWT - Food Science and Technology, 61*, 463-470.

Quirante-Moya, S., García-Ibañez, P., Quirante-Moya, F., Villaño D. & Moreno, D. A. (2020). The role of *Brassica* bioactives on human health: are we studying it the right way? *Molecules, 25*, 1591.

Rangel, M. C., Bertolette, D., Castro, N. P., Klauzinska, M., Cuttitta, F. & Salomon, D. S. (2016). Developmental signaling pathways regulating mammary stem cells and contributing to the etiology of triple-negative breast cancer. *Breast Cancer Research and Treatment, 156*, 211–26.

Russo, M., Spagnuolo, C., Russo, G. L., Skalicka-Wozniak, K., Daglia, M., Sobarzo-Sánchez, E., Nabavi, S. F. & Nabavi, S. M. (2018). Nrf2 targeting by sulforaphane: a potential therapy for cancer treatment. *Critical Reviews in Food Science and Nutrition 58*, 1391–1405.

Saedisomeolia, A., Arzati, M. M., Abdolahi, M., Sedighiyan, M., Rangel, A., Muench, G., Zarezadeh, M., Jafarieh, A. & Honarvar, N. M. (2019). Mechanisms of action of ginger in nuclear factor-kappaB signaling pathways in diabetes. *Journal of Herbal Medicine, 16*, 100239.

Sakurai, H., Morishima, Y., Ishii, Y., Yoshida, K., Nakajima, M., Tsunoda, Y., Hayashi, S.-Y., Kiwamoto, T., Matsuno, Y., Kawaguchi, M., Yamamoto, M. & Hizawa, N. (2018). Sulforaphane ameliorates steroid insensitivity through an Nrf2-dependent pathway in cigarette smoke-exposed asthmatic mice. *Free Radical Biology and Medicine, 129*, 473–485.

Saleh, D. O., Mansour, D. F., Hashad, I. M. & Bakeer, R. M. (2019). Effects of sulforaphane on D-galactose-induced liver aging in rats: Role of keap-1/nrf2 pathway. *European Journal of Pharmacology, 855*, 40–49.

Shah, S. Z. A., Zhao, D. M., Hussain, T. & Yang, L. F. (2017). Role of the AMPK pathway in promoting autophagic flux via modulating mitochondrial dynamics in neurodegenerative diseases: Insight into prion diseases. *Ageing Research Reviews, 40*, 51–63.

Shao, Q., Liu, T., Wang, W., Duan, Q., Liu, T., Xu, L., Huang, G. & Chen, Z. (2020). The Chinese herbal prescription JZ-1 induces autophagy to protect against herpes simplex Virus-2 in human vaginal epithelial cells by inhibiting the PI3K/Akt/mTOR pathway. *Journal of Ethnopharmacology, 254*, 112611.

Silva-Palacios, A., Ostolga-Chavarría, M., Sánchez-Garibay, C., Rojas-Morales, P., Galván-Arzate, S., Buelna-Chontal, M., Pavónf, N., Pedraza-Chaverrí, J., Königsberg, M. & Zazueta, C. (2019). Sulforaphane protects from myocardial ischemia-reperfusion damage through the balanced activation of Nrf2/AhR. *Free Radical Biology and Medicine, 143*, 331–340.

Subedi, L., Cho, K. H., Park, Y. U., Choi, H. J. & Kim, S. Y. (2019). Sulforaphane-enriched broccoli sprouts pretreated by pulsed electric fields reduces neuroinflammation and ameliorates scopolamine-induced amnesia in mouse brain through its antioxidant ability via Nrf2-HO-1 activation. *Oxidative Medicine and Cellular Longevity, 3549274*.

Sun, Y., Zhou, S., Guo, H., Zhang, J., Ma, T., Zheng, Y., Zhang, Z. & Cai, L. (2020). Protective effects of sulforaphane on type 2 diabetes-induced cardiomyopathy via AMPK-mediated activation of lipid metabolic pathways and NRF2 function. *Metabolism Clinical and Experimental, 102*, 154002.

Thangapandiyan, S., Ramesh, M., Miltonprabu, S., Hema, T., Jothi, G. B. & Nandhini, V. (2019). Sulforaphane potentially attenuates arsenic induced nephrotoxicity via the PI3K/Akt/Nrf2 pathway in albino

Wistar rats. *Environmental Science and Pollution Research, 26*, 12247–12263.

Uddin, M. S., Mamun, A. A., Jakaria, M., Thangapandiyan, S. & Ahmad, J. (2020). Emerging promise of sulforaphane-mediated Nrf2 signaling cascade against neurological disorders. *Science of Total Environment, 707*, 135624.

Wang, C., Nie, X., Zhang, Y., Li, T., Mao, J., Liu, X., Gu, Y., Shi, J., Xiao, J., Wan, C. & Wu, Q. (2015). Reactive oxygen species mediate nitric oxide production through ERK/JNK MAPK signaling in HAPI microglia after PFOS exposure. *Toxicology and Applied Pharmacology, 288*, 143–151.

Wang, S., Liu, Z., Wang, L. & Zhang, X. (2009). NF-kappaB signaling pathway, inflammation and colorectal cancer. *Cellular and Molecular Immunology, 6*, 327–334.

Wei, L., Wang, J., Yan, L., Shui, S., Wang, L., Zheng, W., Liu, S., Liu, C. & Zheng, L. (2020). Sulforaphane attenuates 5-fluorouracil induced intestinal injury in mice. *Journal of Functional Foods, 69*, 103965.

Wu, D. M., Zheng, Z. H., Fan, S. H., Zhang, Z. F., Chen, G. Q. & Lu, J. (2020). Sulforaphane administration alleviates diffuse axonal injury (DAI) via regulation signaling pathway of NRF2 and HO-1. *Journal of Cellular Biochemistry, 121*, 430–442.

Wu, W., Peng, G., Yang, F., Zhang, Y., Mu, Z. & Han, X. (2019). Sulforaphane has a therapeutic effect in an atopic dermatitis murine model and activates the Nrf2/HO-1 axis. *Molecular Medicine Reports, 20*, 1761–1771.

Xin, P., Xu, X., Deng, C., Liu, S., Wang, Y., Zhou, X., Ma, H., Wei, D. & Sun, S. (2020). The role of JAK/STAT signaling pathway and its inhibitors in diseases. *International Immunopharmacology, 80*, 106210.

Yan, Z., Gibson, S. A., Buckley, J. A., Qin, H & Benveniste, E. N. (2018). Role of the JAK/STAT signaling pathway in regulation of innate immunity in neuroinflammatory diseases. *Clinical Immunology, 189*, 4–13.

Yang F., Wang F., Liu Y., Wang S., Li X., Huang Y., Xia, Y. and Cao, C. (2018b). Sulforaphane induces autophagy by inhibition of HDAC6-mediated PTEN activation in triple negative breast cancer cells. *Life Sciences, 213*, 149–57.

Yang, G., Yeon, S. H., Lee, H. E., Kang, H. C., Cho, Y. Y., Lee, H. S. & Lee, J. Y. (2018a). Suppression of NLRP3 inflammasome by oral treatment with sulforaphane alleviates acute gouty inflammation. *Rheumatology, 57*, 727–736.

Ye, J. H., Huang, L. Y., Terefe, N. S. & Augustin, M. A. (2019). Fermentation-based biotransformation of glucosinolates, phenolics and sugars in retorted broccoli puree by lactic acid bacteria. *Food Chemistry, 286*, 616–623.

Yi, Y. S. (2018). Regulatory roles of the caspase-11 non-canonical inflammasome in inflammatory diseases. *Immune Network, 18*, 41.

Yu, C., He, Q., Zheng, J., Li, L. Y., Hou, Y. H. & Song, F. Z. (2017). Sulforaphane improves outcomes and slows cerebral ischemic/reperfusion injury via inhibition of NLRP3 inflammasome activation in rats. *International Immunopharmacology, 45*, 74–78.

Zhan, T., Rindtorff, N. & Boutros, M. (2017). Wnt signaling in cancer. *Oncogene, 36*, 1461–1473.

Zhang Z., Wang S., Zhou, S., Yan, X., Wang, Y., Chen, J., Mellen, N., Kong, M., Gu, J., Tan, Y., Zheng, Y. & Cai, L. (2014). Sulforaphane prevents the development of cardiomyopathy in type 2 diabetic mice probably by reversing oxidative stress-induced inhibition of LKB1/AMPK pathway. *Journal of Molecular and Cellular Cardiology, 77*, 42–52.

Zhang, J., Zhang, R., Zhan, Z., Li, X., Zhou, F., Xing, A., Jiang, C., Chen, Y. & An, Li. (2017). Beneficial effects of sulforaphane treatment in Alzheimer's disease may be mediated through reduced HDAC1/3 and increased P75NTR expression. *Frontiers in Aging Neuroscience, 9*, 121.

Zhang, M., An, C., Gao, Y., Leak, R. K., Chen, J. & Zhang, F. (2013). Emerging roles of Nrf2 and phase II antioxidant enzymes in neuroprotection. *Progress in Neurobiology, 100*, 30–47.

Zhang, Y., Gilmour, A., Ahn, Y. H., de la Vega, L. & Dinkova-Kostova, A. T. (2019). The isothiocyanate sulforaphane inhibits mTOR in an NRF2-independent manner, *Phytomedicine, 153062*.

Zhou, A., Hong, Y. & Lv, Y. (2019). Sulforaphane attenuates endometriosis in rat models through inhibiting PI3K/Akt signaling pathway. *Dose-Response: An International Journal*, 1-8.

Zhou, Q., Chen, B., Wang, X., Wu, L., Yang, Y., Cheng, X., Hu, Z., Cai, X., Yang, J., Sun, X., Lu, W., Yan, H., Chen, J., Ye, J., Shen, J. & Cao, P. (2016). Sulforaphane protects against rotenone-induced neurotoxicity in vivo: involvement of the mTOR, Nrf2, and autophagy pathways. *Scientific Reports, 6*, 32206.

In: Fermented Foods
Editor: Oliver Kovalyov
ISBN: 978-1-53618-656-7

Chapter 3

BACTERIOCINOGENIC ACTIVITY OF *LEUCONOSTOC LACTIS* RK18 ISOLATED FROM FERMENTED FOOD

M. Oliveira, J. Barbosa, H. Albano and P. Teixeira*
Universidade Católica Portuguesa,
CBQF - Centro de Biotecnologia e Química Fina –
Laboratório Associado, Escola Superior de Biotecnologia,
Rua Diogo Botelho, Porto, Portugal

ABSTRACT

Safety of fermented foods is mostly ensured by the presence of lactic acid bacteria, which besides having an important role in the enhancement of food shelf-life, also enhance the organoleptic characteristics of the products. One of the major attributes of lactic acid bacteria is their bactericidal activity against spoilage and pathogenic bacteria by the production of proteinaceous substances that are commonly known as bacteriocins.

* Corresponding Author's Email: pcteixeira@porto.ucp.pt.

The present study aimed to characterize a proteinaceous compound produced by *Leuconostoc lactis* RK18, isolated from a fermented Cambodian food product.

It was found that *Ln. lactis* RK18 produced a Class IIa bacteriocin, capable of inhibiting important foodborne pathogens such as *Listeria monocytogenes*. Bacteriocin RK18 remained stable in the temperature range between 30°C to 80°C) and at pH values ranging from 4 to 6. Although susceptible to some detergents, it showed great resistance to several enzymes. RK18 was produced in the early stage of *Ln. lactis* RK18 growth cycle, which brings an advantage when competitive growth is desired or for pathogen inhibition purposes, since it can inhibit pathogenic bacteria before its exponential growth.

Even though more studies must be performed, *Ln. lactis* RK18 appears to have the potential to play an important role in the safety of food products and could be a good candidate to be used as a starter culture in the production of other fermented products.

Keywords: antimicrobial activity, fermented food products, lactic acid bacteria, bacteriocin

INTRODUCTION

Nearly five thousand people perish each year due to the consumption of contaminated foods (WHO, 2017). Foodborne outbreaks are progressively increasing but the real number of cases is still unknown. According to Centers for Disease Control and Prevention (CDC), outbreaks related to the consumption of fermented foods such as soft cheese and deli-sliced meats and cheeses have been associated with many pathogenic bacteria including *Listeria monocytogenes* (CDC, 2019). *Listeria monocytogenes* causes listeriosis, a rare but potentially fatal foodborne infection with increasing incidence in the European Union (EFSA, 2019). *Listeria monocytogenes* is ubiquitous in the environment being found in water, soil and faeces, granting the possibility of contamination in food-processing environments. It has remarkable capability to evade host immunity system and survive in the human body, crossing the intestinal and blood-brain barrier, thereby being life-threatening and causing severe infections such as encephalitis and meningitis to an immune-compromised host (Jordan and McAuliffe, 2018).

Fermented foods produced through the activity of lactic acid bacteria (LAB) are fashionable due to the health benefits attributed to their consumption (Marco et al., 2017). Given their involvement in commercial food fermentations, LAB comprise the most explored microorganisms concerning their physiology, biochemistry, genetics and evolution proving to be the most "domesticated" group of bacteria (Douglas et al., 2010). Lactic acid bacteria are well adapted to survive, proliferate and exert their functional roles in food environments (Papadimitriou et al., 2016). Bactericidal activity against spoilage and pathogenic bacteria, achieved by the acidification of food matrices during lactic acid fermentation, is one of the major attributes of LAB when used as starter cultures (Papadimitriou et al., 2016). Since LAB are relatively acid tolerant, acidification of the medium can cause a stress to pathogenic bacteria, damaging both the cell wall and the cell membrane. This allows the action of exerted lactic acid that is viewed as an antimicrobial agent against competing microorganisms (Even et al., 2002).

Leuconostoc is among the most common genera of LAB present in fermented food products (Papadimitriou et al., 2016; Rezac et al., 2018). *Leuconostoc* spp. are ubiquitous microorganisms, especially found in fermented meats, vegetables, and dairy products, but also in plant-matter and human clinical sources (Liu, 2016). Leuconostocs are able to contribute to human health improvement by producing and excreting certain vitamins and antimicrobial compounds, e.g., bacteriocins, active against spoilage and pathogenic microorganisms. Most studies have investigated the contribution of leuconostocs to the enhancement of nutritional value and the quality of fermented foods (Liu, 2016; Hwang et al., 2018). Despite the association of human infections to *Leuconostoc* species, this is related to opportunistic strains that affect susceptible immunocompromised individuals and, overall, the genus is Generally Recognized As Safe (GRAS) (Liu, 2016).

It has been known that different organisms, such as animals, plants, insects and bacteria, can produce antibacterial substances, which includes organic acids, antibiotics and bacteriocins (Ray and Joshi, 2014; Yang et al., 2014). Bacteriocins are defined as a proteinaceous inhibitors or antimicrobial peptides that are produced by bacteria. In general, these compounds act as antibacterial through the inhibition of target cell wall synthesis or through depolarization of the cell membrane (Ray and Joshi, 2014; Yang et al., 2014). Despite the killing ability of bacteriocins, these substances are sensitive to proteases meaning that, in general, are harmless to the human body and the environment. Lactic acid bacteria are known to be the most studied bacteriocin producers, giving the high variety of bacteriocins of different structures, sizes, physical and chemical properties and inhibitory spectrum (Yang et al., 2014).

This study aimed to characterize an anti-listerial bacteriocin, RK18, produced by *Leuconostoc lactis* RK18 isolated from a fermented Cambodian food product.

METHODS

Bacterial Strains, Media and Growth Conditions

Leuconostoc lactis RK18 used in this study, isolated from a Cambodian fermented food product (Khmer Fermented Rice Fresh Noodles), was grown in de Man Rogosa & Sharp agar (MRS, Biokar, Beauvais, France) at 30°C for 48 - 72 h. For an initial screening, seven microorganisms were used as target strains – four strains of *L. monocytogenes* (each strain belonging to a different serogroup), *Enterococcus faecalis* ATCC 29212, *Staphylococcus aureus* ATCC 25213 and *Enterococcus faecium* DSMZ 13590 (Table 1). These were cultured in TSAYE (Tryptic Soy Agar (TSA, Biokar) with 6 g/l of Yeast Extract (YE, Lab M, Lancashire, United Kingdom)) and incubated at 37°C for 24 hours. All strains were stored at -20°C in the presence of 30% (v/v) glycerol (Sigma, Steinheim, Germany) and sub-cultured twice before use in assays.

Table 1. Origin of the target organisms used in this study

Microorganisms	Species	Source
Gram-positive	*Bacillus cereus*	ESB culture collection
	Bacillus subtilis	
	Bacillus stearothermophilus	
	Listeria monocytogenes SCOTT A	
	[a]*Listeria innocua* 2030c	
	Staphylococcus aureus 18N (Methicillin-resistant *Staphylococcus aureus*-MRSA)	
	Staphylococcus aureus 2037 M1 (Methicillin-sensitive *S. aureus*-MSSA)	
	Enterococcus faecalis ATCC 29212*	ATCC
	Staphylococcus aureus ATCC 25213*	
	Enterococcus casseliflavus DSMZ 20680	DSMZ
	Enterococcus faecalis DSMZ 12956	
	Enterococcus faecium DSMZ 13590*	
	Enterococcus flavescens DSMZ 7370	
	Enterococcus gallinarum DSMZ 20628	
	Listeria monocytogenes L 7946	McLauchlin et al. (1997)
	Listeria monocytogenes L 7947	
	Listeria monocytogenes NCTC 1194 (serogroup 4b)*	LRCESB
	Listeria monocytogenes CECT 911 (serogroup 1/2c)*	
	Listeria monocytogenes CECT 936 (serogroup 1/2b)*	
	Listeria monocytogenes CEP 104794 (serogroup 1/2a)*	
Gram-negative	*Acinetobacter baumanii* R	ESB culture collection
	Acinetobacter baumanii S-1	
	Acinetobacter baumanii S-2	
	Acinetobacter calcoaceticus S	
	Acinetobacter calcoaceticus R	
	Klebsiella pneumoniae	
	Proteus mirabilis	
	Proteus vulgaris	
	Pseudomonas aeruginosa	
	Salmonella Braenderup	
	Salmonella Enteritidis	
	Salmonella Enteritidis 417536	
	Salmonella Enteritidis 545047	
	Salmonella Tiphymurium	
	Yersinia enterocolitica	
	Escherichia coli ATCC 25922	ATCC
	Yersinia enterocolitica NCTC 10406	NCTC

ESB – organisms deposited on the culture collection of Escola Superior de Biotecnologia; DSMZ – German Collection of Microorganisms and Cell Culture; LRCESB – isolates deposited on the Culture Collection of *Listeria* Research Center of Escola Superior de Biotecnologia; ATCC – American Type Culture Collection; NCTC – National Collection of Types cultures – Culture Collection of Public Health England. S – sensitive to several tested antibiotics; R – resistant to several tested antibiotics. *Target organisms used in the initial screening. [a]Public Health Laboratory Service (PHLS), London

Study of Antimicrobial Activity of *Ln. Lactis* RK18

Antimicrobial Activity Screening

Antibacterial activity was tested by the agar spot test (Van Reenen et al., 1998). One colony of *Ln. Lactis* RK18 grown on TSAYE was transferred to 10 ml of TSBYE (Tryptic Soy Broth (TSB, Biokar) with 6 g/l of YE) and incubated overnight at 37°C. Suspensions of each target bacterial culture were spread onto TSAYE plates and 10 μl drops of the LAB culture, grown twice in MRS broth (Biokar), were spotted on the lawns of target organisms and incubated overnight at 37°C. Inhibition was recorded as positive if a translucent halo zone was observed around the spot. *Pediococcus acidilactici* HA-6111-2 (from Culture Collection of *Escola Superior de Biotecnologia*) was used as anti-listerial control strain (Albano et al., 2009).

Antimicrobial Activity Characterization

The nature of the inhibition was determined by the qualitative agar-diffusion technique according to Tomé et al. (2006). Culture broth of *Ln. Lactis* RK18 was centrifuged at 7000 rpm for 10 min at 4°C (Centrifuge 5427 R, Eppendorf, Hamburg, Germany), the clear supernatants were sterilized by membrane filtration (0.2 μm; Sartorius, Goettingen, Germany) and their pH adjusted to ≈ 5.0 and 6.0 with sterilized solution of sodium hydroxide (1M NaOH, José M. Vaz Pereira, Lisbon, Portugal). To determine whether the inhibition was due to hydrogen peroxide production or to proteinaceous compounds, neutralized supernatants were treated with catalase (500 IU/ml; Sigma) and proteinase K (0.1 mg/ml, sterile; Sigma), respectively, for 1 h at 37ºC. Culture grown (C), cell free supernatant (CFS), neutralized cell-free supernatant (CFSn), neutralized cell-free supernatant treated with catalase (CFSnC) and neutralized cell-free supernatant treated with proteinase K (CFSnK), were spotted against the target microorganisms. *Pediococcus acidilactici* HA-6111-2 was used as anti-listerial control strain (Albano et al., 2009). Inhibition by a proteinaceous substance (possible bacteriocin production) was assumed if a translucent halo zone was observed around all the spots except for cell-free supernatant treated with proteinase K.

Maximum Bacteriocin Production (AU/ml) during Growth of Ln. lactis RK18

To determine de maximum bacteriocin production during its growth, 1% (v/v) of an overnight *Ln. lactis* RK18 culture was inoculated in MRS broth (100 ml) and incubated at 30ºC. Changes in pH and Optical Density (O.D., 600 nm) were recorded every hour during 48 h. Every three hours, aliquots of 10 ml and 1 ml were taken to test bacteriocin activity (AU/ml) in the treated cell-free supernatant against strains of *L. monocytogenes* (Van Reenen et al., 1998) and to determine viable cell counts (CFU/ml) of *Ln. lactis* RK18.

To determine the bacteriocin activity (AU/ml), cells from 10 ml aliquots were harvest by centrifugation (7000 rpm, 10 min, 4ºC; Centrifuge 5427 R, Eppendorf) and the cell-free supernatants (CFS) adjusted to pH 5.0-6.0 with 1M NaOH and incubated at 80ºC for 10 minutes (CFSnh) (Van Reenen et al., 1998). Then, each treated CFS was successive diluted on PBS (VWR Chemicals, Ohio, USA) and 10 µl aliquots of each dilution were spotted onto a soft agar plate (BHI with 0.7% w/v agar) seeded with approximately 10^6 CFU/ml of each target *L. monocytogenes* strain. Plates were incubated at 30ºC during 24 h to 48 h. *Pediococcus acidilactici* HA-6111-2 was used as control. Antimicrobial activity was expressed as arbitrary units (AU) per ml and according to Van Reenen et al. (1998) one AU is defined as the reciprocal of the highest dilution showing a clear zone of growth inhibition.

Effect of pH, Enzymes, Temperature, Detergents, Surfactants and Protease Inhibitors on Bacteriocin Activity

The evaluation of the following tests was performed on treated cell-free supernatant (CFS) of *Ln. lactis* RK18 prepared as described in the previous section.

The effect of enzymes proteinase K, papain, pepsin and catalase (all from Sigma) was determined by incubating 1 ml of each CFS neutralized and heated at 80°C (CFSnh) for 2 hours in the presence of both 1 mg/ml and 0.1 mg/ml of each enzyme.

The effect of temperature on bacteriocin activity was tested by incubating CFSn at 4, 25, 30, 60, 80 and 100°C for 1 h and 2 h. Bacteriocin activity was also tested after 15 min at 121°C.

For the evaluation of the effect of pH on the bacteriocin activity, CFS was adjusted from pH 2.0 to 12.0 (at increments of two pH units) with sterile 1M NaOH or 1M Hydrochloric Acid (HCl). After 1 h of incubation at room temperature, the samples were readjusted to pH 5 to 6 and incubated at 80°C for 10 min.

Finally, 1% (w/v) Tween 20, Tween 80, Triton X-100 (Bio-Rad Laboratories, Richmond, CA, USA), EDTA (Panreac Quimica SA, Barcelona, Spain), Ox-bile (Pronadisa), Urea (Merck, Darmstadt, Germany), SDS, NaCl, Sodium Carbonate and Sodium deoxycholate (all for Sigma) were added to the cell-free supernatant neutralized CFSn. Untreated CFS and detergents in water at the same concentrations were used as controls. All samples were incubated at 30°C for 5 h.

For all the conditions, antimicrobial activity was monitored by agar-spot test method (Van Reenen et al., 1998) and the 4 serogroups of *L. monocytogenes*, mentioned in the previous section, were used as target strains.

Cell Lysis of Target Microorganisms

Twenty percent (v/v) of bacteriocin-containing cell-free supernatant, neutralized and filtered, was added to 100 ml cultures of each target organism (when reached early exponential stages). Every hour, optical density at 600 nm was read for 12 h. Viable cell counts (CFU/ml) were determined every 2 h between 0 (time at which inoculum of each of the 4 serogroups of *L. monocytogenes* strain was added at flasks containing TSBYE) and 12 h, by plating onto TSAYE and incubated at 37°C for 24 h. Cell suspensions of each target without added bacteriocin served as control.

Adsorption Studies and Partial Purification of Ln. Lactis RK18 Bacteriocin

Adsorption of bacteriocin RK18 was conducted according to the method described by Yang et al. (1992). After pH adjusted to 6.0, bacteriocin producing cells were harvested, washed with sterile 0.1 M phosphate buffer (pH 6.5), resuspended in 10 ml of 100 mM NaCl (pH 2.0) and agitated for 1 h at 4°C, to allow delaminating bacteriocin from the cells. Then, cells were harvested, and cell-free supernatant was neutralized and tested for bacteriocin activity as described by Van Reenen et al. (1998).

The supernatant resulting from the first centrifugation was kept at 4°C for partial purification. Later, ammonium sulphate was added gradually to the supernatant stored to reach 40%, 60% and 80% of saturation and each solution was kept at slow stirring during 4 h at 4°C. After centrifugation (12000 rpm, 20 min, 4°C) precipitated proteins in the pellet and floating on the surface were collected and dissolved in 25 mM ammonium acetate buffer (pH 6.5) following the method described by Sambrook et al. (1989). All samples were stored at -20°C.

Molecular Size of Ln. Lactis RK18 Bacteriocin

The samples collected and stored at -20°C, as described in the previous section, were separated by tricine-SDS-PAGE according to Schägger and Von Jagow (1987). A low molecular weight marker with sizes ranging from 6.5 kDa to 270 kDa (Grisp) was used.

After running, the gel was split in two. One half of the gels was fixed with 20% isopropanol and 10% acetic acid, and the other half was stained with Coomassie Brilliant Blue R250 (Bio-Rad) to visualize the position of the peptide band. The position of the active bacteriocin was determined by overlaying the other half of the gel (not stained and extensively pre-washed with the sterile distilled water) with cells of *L. monocytogenes* CEP 104794 (10^6 CFU/ml), firstly embedded in BHI agar (0.7% agar w/v; Biokar).

Bacteriocin Activity Spectrum

Antimicrobial activity due to bacteriocin production by *Ln. lactis* RK18 was screened against 20 Gram-positive and 17 Gram-negative bacteria (Table 1) according to the method described by Van Reenen et al. (1998). Cell-free supernatant with antimicrobial activity was treated with proteinase K (0.1 mg/ml, sterile; Sigma) to confirm that the activity was caused by the presence of the bacteriocin. The confirmation of antimicrobial activity was assumed if a translucent halo zone was observed around the spots.

RESULTS AND DISCUSSION

Inhibitory activity can be achieved by competition against microorganisms or by compounds that are produced and secreted by some bacteria. So, using this initial screening method, the observation of an inhibition zone, may result from competition, lactic acid production with consequent pH decrease, hydrogen peroxide or bacteriocin production. *Leuconostoc lactis* RK18 demonstrated antimicrobial activity due to bacteriocinogenic activity against all four *Listeria* strains and *E. faecalis* ATCC 29212 tested. The effectiveness of LAB in inhibiting Gram-positive bacteria is widely recognized and described by several authors (Albano et al., 2007; Albano et al., 2009; Abrams et al., 2011; Peng et al., 2017).

The correlation between growth of *Ln. lactis* RK18 and its anti-listerial activity (AU/ml) was also determined. The antimicrobial activity (AU/ml) of the bacteriocin produced by *Ln. lactis* RK18 was evaluated against strains of *L. monocytogenes* belonging to four serogroups (Figure 1).

The higher activity of bacteriocin was observed against *L. monocytogenes* CECT 936 (L3) with 12800 AU/ml during growth of *Ln. lactis* RK18 for 6 h-12 h (Figure 1). For *L. monocytogenes* NCTC 1194 (L1) and *L. monocytogenes* CEP 104794 (L4) the maximum activity of the bacteriocin (6400 AU/ml) was reached after 6h and 18h, respectively. The lowest bacteriocin activity (800 AU/ml) was observed during the period of 12 h to 18 h against *L. monocytogenes* CECT 911 (L2). This shows that the activity of the bacteriocin produced by *Ln. lactis* RK18 is strain-dependent. Also, the changes in pH values decreased from 6.5 (in the beginning of the assay) to 4.6 (at the end of the screening) reaching its minimum (pH 4.4) after 15 h of growth. The number of viable cells of *Ln. lactis* RK18 increased approximately, ca. 2.6 log (maximum cell growth at 9 h of growth). Afterwards, pH values reached a plateau and cell viability decreased 2 logs. Low levels of bacteriocin activity were recorded after 24 h of growth decreasing gradually until 200 AU/ml against *L. monocytogenes* NCTC 1194 (L1) and *L. monocytogenes* CECT 936 (L3). The peptide lost activity against *L. monocytogenes* CECT 911 (L2) and *L. monocytogenes* CEP 104794 (L4) after 24 h and 30 h, respectively.

For the same inoculum concentrations as used in this study (1% v/v), a similar behavior was previously observed for the bacteriocinogenic *Pediococcus pentosaceus* K34, with bacteriocin production reaching maximum values of 12800 AU/ml against *L. monocytogenes* (Abrams et al., 2011). However, these results are somewhat different from the results reported by Hwang et al. (2018), since *Ln. lactis* SD501 demonstrated maximum activity (735 AU/ml) at 9 h of growth against *L. monocytogenes* ATCC 19114. This confirms that bacteriocin production by probiotic strains is strain-specific. Similarly, this bacteriocin activity decreased gradually after prolonged incubation. Nevertheless, similar maximum activity values against other *L. monocytogenes* strains were reported to other bacteriocin-producing *Leuconostoc* strains and to other LAB closely related to this genus (Héchard et al., 1992; Arakawa et al., 2016; Chen et al., 2018; Lasik-Kurdyś and Sip, 2019). The reduction of antimicrobial activity of bacteria may be due to various factors, including auto-acidification after prolonged incubation, proteolytic degradation by extracellular proteases or protein aggregation (Papagianni and Papamichael, 2011; Papadimitriou et al., 2016).

Leuconostoc strains generally produce bacteriocins in early or in late stages of their growth cycle (Stiles, 1994). In this study, it was observed that *Ln. lactis* RK18 produced bacteriocin in early stages of its growth cycle, which brings an advantage in circumstances when competitive growth is desired or for pathogen inhibition purposes, since it can inhibit pathogenic bacteria before they enter the exponential growth phase.

Different physicochemical treatments may affect antimicrobial activity of bacteriocins (Gharsallaoui et al., 2016). The stability of the RK18 bacteriocin when exposed to different treatments is presented in Tables 2, 3 and 4.

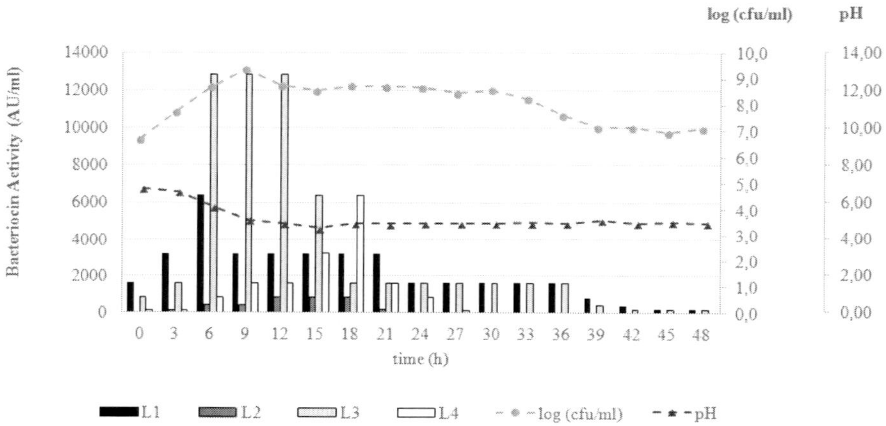

Figure 1. Production of bacteriocin by *Ln. lactis* RK18 in MRS broth at 30°C. Antimicrobial activity of treated cell-free supernatant is presented as AU/ml against *Listeria monocytogenes* of four serogroups: *L. monocytogenes* NCTC 1194 – 4b (**L1**); *L. monocytogenes* CECT 911 – 1/2c (**L2**); *L. monocytogenes* CECT 936 – 1/2b (**L3**) and *L. monocytogenes* CEP 104794 – 1/2 a (**L4**). Viable cell counts of *Ln. lactis* RK18 are presented as log (CFU/ml) (-●-) and pH changes are also represented (-▲-).

The antimicrobial activity of bacteriocin RK18 was affected (25% - 100% of reduction) at pH values below 2.0 and above 8.0, suggesting that this peptide is sensitive to acidic and alkaline conditions (Table 2). The bacteriocin remained stable after incubation for 1 h at pH 4.0 and 6.0 (0-6% of reduction). Even though the negative effect of pH on the antimicrobial activity, bacteriocin still retained its high anti-listerial ability against strains of serogroups 4b (L1) and 1/2c (L2). The acidification can cause solubilization of metals in some LAB, which may be toxic, depending on their concentration, and low values of pH can result in the denaturation of proteins (Papadimitriou et al., 2016), corroborating the stated data. The stability of bacteriocin activity at different pH values was also reported for other *Leuconostoc* strains (Arakawa et al., 2016; Bello et al., 2018; Hwang et al., 2018) and to other LAB (Albano et al., 2007).

Table 2. Effect of pH, detergents, surfactants and protease inhibitors on the antimicrobial activity of bacteriocin RK18 (expressed in percentage values) against *Listeria monocytogenes* belonging to four serogroups

		L1	L2	L3	L4
pH	2	50%	100%	13%	25%
	4	0%	0%	6%	3%
	6	0%	0%	6%	6%
	8	100%	25%	6%	25%
	10	100%	100%	25%	25%
	12	100%	13%	3%	3%
Detergents	Tween 20	0%	0%	100%	75%
	Tween 80	0%	0%	100%	50%
	Triton X-100	50%	75%	100%	50%
	SDS	50%	75%	100%	88%
	EDTA	50%	0%	100%	75%
	Ox-Bile	97%	88%	100%	97%
	Urea	50%	50%	50%	75%
	NaCl	0%	0%	25%	25%
	Sodium carbonate	50%	0%	94%	75%
	Sodium deoxycholate	50%	50%	75%	88%

(L1) *L. monocytogenes* NCTC 1194 – 4b; (L2) *L. monocytogenes* CECT 911 – 1/2c; (L3) *L. monocytogenes* CECT 936 – 1/2b and (L4) *L. monocytogenes* CEP 104794 - 1/2a.

Bacteriocin RK18 was sensitive to all the detergents tested but remained stable against *L. monocytogenes* CECT 911 after treatment with surfactants (Tween-20 and Tween-80), EDTA and salts (NaCl and sodium deoxycholate). Similar results were observed against *L. monocytogenes* NCTC 1194, since bacteriocin remained stable after treatment with surfactants and NaCl. However, almost a total loss of activity against the other strains tested was observed after treatment with all the detergents. This could indicate a high sensitivity of the bacteriocin to protease inhibitors, corroborating the reported data for other LAB species (Benmouna et al., 2018).

Table 3. Reduction of antimicrobial activity of bacteriocin RK18 (expressed in percentage values) against four serogroups of *Listeria monocytogenes*, under the effect of temperature

		L1		L2		L3		L4	
		1h	2h	1h	2h	1h	2h	1h	2h
T (°C)	4	0%	50%	0%	50%	94%	97%	88%	94%
	25	0%	50%	0%	75%	88%	94%	75%	88%
	30	0%	0%	0%	0%	0%	0%	0%	0%
	37	0%	0%	0%	0%	0%	0%	0%	0%
	60	0%	0%	0%	0%	0%	0%	0%	0%
	80	0%	0%	0%	0%	0%	0%	0%	0%
	100	50%	75%	50%	50%	88%	94%	88%	88%
	121	98%	98%	100%	100%	100%	100%	100%	100%

(L1) *L. monocytogenes* NCTC 1194 – 4b; (L2) *L. monocytogenes* CECT 911 – 1/2c; (L3) *L. monocytogenes* CECT 936 – 1/2b and (L4) *L. monocytogenes* CEP 104794 - 1/2a.

Dissimilar results were reported for other *Leuconostoc* strains (Bello et al., 2018) and other LAB (Albano et al., 2007), showing that detergents, surfactants and salts had no effect on the activity of the antimicrobial susbtances produced.

Bacteriocin RK18 was shown to be thermostable, since its activity was maintained at temperatures ranging from 30°C to 80°C (Table 3). Residual activity was observed at lower (below 30°C) and at higher temperatures (at boiling and sterilization temperatures). Similar results were observed for other thermo-stable bacteriocins (Albano et al., 2007; Arakawa et al., 2016; Bello et al., 2018; Hwang et al., 2018).

Complete inactivation of antimicrobial activity was observed after treatment with the proteolytic enzyme, proteinase K, confirming its proteinaceous nature (Table 4). Overall, no reduction of activity was observed after treatment with the other proteases tested (papain and pepsin) and after treatment with catalase. Resistance to catalase suggests that carbohydrates are not bound to the enzyme and its antimicrobial effect was not derived from hydrogen peroxide (H_2O_2). However, the concentration of enzymes seems to be an important factor, since there was a partial reduction (50% for catalase at 1 mg/ml) or almost total reduction of activity (ca. 88% and 75% of reduction for protease enzymes at 1 mg/ml and 0.1 mg/ml, respectively) against two *L. monocytogenes* strains (L2 and

L3). Similar results, were observed for other *Leuconostoc* strains (Arakawa et al., 2016; Chen et al., 2018; Bello et al., 2018; Hwang et al., 2018) and for other LAB (Albano et al., 2007).

Overall, bacteriocin RK18 demonstrated high sensitivity to detergents, surfactants and protease inhibitors, but it remained stable when subjected to more acidic pH, keeping its antimicrobial activity against some *L. monocytogenes* strains. It was also demonstrated that the antimicrobial activity of the bacteriocin was different against the *L. monocytogenes* strains tested. This highlights the importance to perform this kind of studies for more than one target strain.

The effect of bacteriocin RK18, at its maximum activity on the growth of *L. monocytogenes* is presented in Figure 2. When the supernatant of *Ln. lactis* RK18 was added to a mid-log culture (4 h-old) of *L. monocytogenes* strains, a decrease of approximately 2 log cycles was observed for *L. monocytogenes* NCTC 1194 (L1) and *L. monocytogenes* CECT 911 (L2). Similarly, the addition of the proteinaceous substance (maximum activity of 12800 AU/ml against L3 and maximum activity 6400 AU/ml against L4, observed previously) to a mid-log (5 h-old) culture of *L. monocytogenes* CECT 936 (L3) and *L. monocytogenes* CEP 104794 (L4) cell growth was repressed in a similar way. However, after 7 hours it was observed a gradual growth of all *L. monocytogenes* until 11 h of incubation. The bacteriocin seems to have effect on the *L. monocytogenes* strain viability at the moment of addition, but this effect appears to be attenuated along the time. Therefore, *Ln. lactis* RK18 seemed to be bacteriostatic to exponential-phase cells of all *L. monocytogenes* strains investigated. No changes in cell numbers of all *L. monocytogenes* were recorded for the untreated (control) samples.

Similar results were observed for other LAB strains closely related to the *Leuconostoc* genus where the addition of bacteriocin to a mid-log phase, resulted in a significant decline in the growth curve of the pathogenic target strains (Albano et al., 2007; Abrams et al., 2011; Lasik-Kurdyś and Sip, 2019).

Table 4. Reduction of antimicrobial activity of bacteriocin RK18 (expressed in percentage values) against four serogroups of *Listeria monocytogenes*, under the effect of different enzymes, at two concentrations

Enzymes	L1		L2		L3		L4	
	1mg/ ml	0.1mg /ml	1mg/ ml	0.1mg/ ml	1mg/ ml	0.1mg/ ml	1mg/ ml	0.1mg/ ml
Proteinase K	100%	100%	100%	100%	100%	100%	100%	100%
Papain	0%	0%	0%	0%	88%	75%	0%	0%
Pepsin	0%	0%	50%	0%	88%	75%	0%	0%
Catalase	0%	0%	50%	0%	50%	0%	0%	0%

(L1) *L. monocytogenes* NCTC 1194 – 4b; (L2) *L. monocytogenes* CECT 911 – 1/2c; (L3) *L. monocytogenes* CECT 936 – 1/2b and (L4) *L. monocytogenes* CEP 104794 - 1/2a.

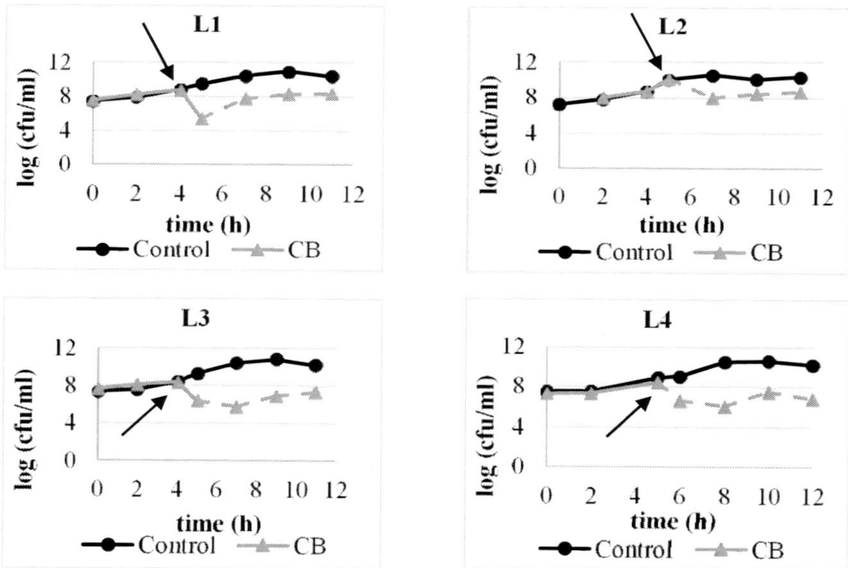

Figure 2. Effect of bacteriocin on growth of *Listeria monocytogenes* belonging to four serogroups: *L. monocytogenes* NCTC 1194 (L1); *L. monocytogenes* CECT 911 (L2); *L. monocytogenes* CECT 936 (L3) and *L. monocytogenes* CEP 104794 (L4). Viable cell counts of each target are presented as log (CFU/ml). Control represents target cultures without added bacteriocins (black); CB represents target cultures with added bacteriocins (grey). Dotted line means that isolate was reduced to values below the detection limit of the enumeration technique.

Regarding adsorption and partial purification, bacteriocin RK18 activity was not detected after treatment of *Ln. lactis* RK18 with 100 mM NaCl at pH 2.0 (0 AU/ml), suggesting that the bacteriocin did not adhere to the surface of the producer cells. Similar results were reported for pediocin HA-6111-2 (Albano et al., 2007) and for the bacteriocin produced by bacPPK34 (Abrams et al., 2011).

As observed in Figure 3, the molecular size of RK18 bacteriocin was below 6.5 kDa. It is also possible to observe the position of the active bacteriocin by a zone of growth inhibition against *L. monocytogenes* CEP 104794, which was firstly embedded in BHI soft agar. This zone of growth inhibition is coincident with the position of the active bacteriocin in the molecular mass marker.

This small peptide is within the range of most bacteriocins reported for the genus *Leuconostoc* and for bacteriocins belonging to class II. Hwang et al. (2018) observed a single low molecular weight band of *Ln. lactis* SD501 (7kDa) and other authors have stated the classification of class IIa bacteriocins (known to be mainly anti-listerial proteinaceous substances) within *Leuconostoc* genus (Héchard et al., 1992; Arakawa et al., 2016; Chen et al., 2018; Hwang et al., 2018).

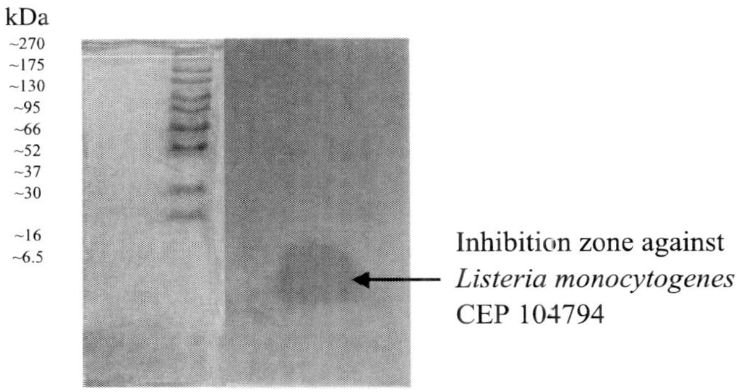

Figure 3. Tricine-SDS PAGE of RK18 bacteriocin. Lane 1: peptide bands stained with Coomassie Blue R250 (60% ammonium sulphate saturated); lane 2: molecular mass marker; lane 3: zone of growth inhibition against *Listeria monocytogenes* CEP 104794.

The anti-listerial activity, wide pH resistance and thermostability suggest that the bacteriocin produced by *Ln. lactis* RK18 may be classified as a class IIa bacteriocin (Yang et al., 2014). Genetic studies on the expression of the gene encoding bacteriocin production, amino acid sequence and mode of action will have to be done to confirm this hypothesis. Other authors also found class IIa bacteriocins produced by *Leuconostoc* species, such as mesentericin Y105 produced by *Ln. mesenteroides* ssp. *mesenteroides* (Stiles, 1994).

The broad spectrum of antimicrobial activity against Gram-positive bacteria is characteristic of many class IIa bacteriocins and the activity against *L. monocytogenes* is particularly relevant (Peng et al., 2017; Bello et al., 2018). The inhibitory activity of bacteriocins may vary given that some only inhibit taxonomically related Gram-positive bacteria, and other are active against a broader range of Gram-positive and Gram-negative microorganisms (Harzallah and Belhadj, 2013).

Bacteriocin RK18 inhibited the growth of all the strains of *L. monocytogenes* investigated. Inhibitory ability against other Gram-positive microorganisms such as *E. faecalis* ATCC 29212 as well as one Gram-negative microorganism, *Yersinia enterocolitica* NCTC 10406 was also observed. This activity spectrum showed that bacteriocin produced by *Ln. lactis* RK18 has an important anti-listerial activity, but also activity against other important foodborne pathogens. Hwang et al. (2018) also reported antimicrobial activity of *Ln. lactis* SD501 against *L. monocytogenes* ATCC 19114 and *E. faecalis* (VRE) CCARM 0011, corroborating the data observed. This is an important advantage that contributes to turn the use of this microorganism even more appealing in the food industry.

Overall, *Ln. lactis* RK18 produced a Class IIa bacteriocin, capable of inhibiting important foodborne pathogens such as *L. monocytogenes*, stable at temperatures ranging from 30ºC to 80ºC and at pH values ranging from 4 to 6, and although susceptible to some detergents, it showed high resistance to several enzymes. Although further experiments are required, this strain appears to be a potential bioprotective culture with application in the food industry.

Acknowledgments

This work was supported by project "Biological tools for adding and defending value in key agro-food chains (bio – n2 – value)", no NORTE-01-0145-FEDER-000030, funded by Fundo Europeu de Desenvolvimento Regional (FEDER), under Programa Operacional Regional do Norte - Norte2020. The authors would also like to thank the scientific collaboration under the FCT project UID/Multi/50016/2019. Financial support for author J. Barbosa was provided by a post-doctoral fellowship SFRH/BPD/113303/2015 (FCT).

References

Abrams, Daniel, Barbosa, Joana, Albano, Helena, Silva, Joana, Gibbs, Paul A; Paula Teixeira. "Characterization of bacPPK34 a bacteriocin produced by *Pediococcus pentosaceus* strain K34 isolated from "Alheira"." *Food Control*, 22, no. 6, (2011), 940-946. https://doi.org/10.1016/j.foodcont.2010.11.029.

Albano, Helena, Pinho, Catarina, Leite, Daniela, Barbosa, Joana, Silva, Joana, Carneiro, Luísa, Magalhães, Rui, Hogg, Tim; Paula Teixeira. "Evaluation of a bacteriocin-producing strain of *Pediococcus acidilactici* as a biopreservative for "Alheira", a fermented meat sausage." *Food Control* 20, no 8, (2009), 764-770. https://doi.org/10.1016/j.foodcont.2008.09.021.

Albano, Helena, Todorov, Svetoslav D; van Reenen, Carol A; Hogg, Tim, Dicks, Leon M; Paula Teixeira. "Characterization of two bacteriocins produced by *Pediococcus acidilactici* isolated from "Alheira", a fermented sausage traditionally produced in Portugal." *International Journal of Food Microbiology*, 116, no. 2, (2007), 239-247. https://doi.org/10.1016/j.ijfoodmicro.2007.01.011.

Arakawa, Kensuke, Yoshida, Saki, Aikawa, Hiroki, Hano, Chihiro, Bolormaa, Tsognemekh, Burenjargal, Sedkhuu, Taku Miyamoto.

"Production of a bacteriocin-like inhibitory substance by *Leuconostoc mesenteroides* subsp. *dextranicum* 213M0 isolated from Mongolian fermented mare milk, airag." *Animal Science Journal*, 87, no. 3, (2016), 449456. https://doi.org/10.1111/asj.12445.

Bello, Olorunjuwon, Babalola, Olubukola Oluranti, Adegboye, Mobolaji, Fashola, Muibat O; Temitope K. Bello. "Partial purification, characterization and application of bacteriocin from bacteria isolated *Parkia biglobosa* seeds." *Natural and Engineering Sciences*, 3, no. 2, (2018), 72-94. https://doi.org/10.28978/nesciences.424517.

Benmouna, Zahia, Dalache, Fatiha, Karam, Noureddine, Halima Zadi-Karam. "Optimization and some characteristics of bacteriocin produced by *Enterococcus* sp. CM9 collected from Mauritanian camel milk." *Emirates Journal of Food and Agriculture*, 30, no. 4, (2018), 275-282. https://doi.org/10.9755/ejfa.2018.v30.i4.1662.

Centers for Disease Control and Prevention (CDC). "List of Selected Multistate Foodborne Outbreak Investigations", 2019, accessed June 1, 2020, https://www.cdc.gov/foodsafety/outbreaks/multistate-outbreaks/outbreaks-list.html.

Chen, Yi-Sheng, Wu, Hui-Chung, Kuo, Chen-Yu, Chen, Yu-Wei, Fujitoshi Yanagida. "Leucocin C-607, a novel bacteriocin from the multiple-bacteriocin-producing *Leuconostoc pseudomesenteroides* 607 isolated from persimmon." *Probiotics and Antimicrobial Proteins*, 10, no. 2, (2018), 148-156. https://doi.org/10.1007/s12602-017-9359-6.

Douglas, Grace L; Todd R. Klaenhammer. "Genomic evolution of domesticated microorganisms." *Annual Review of Food Science and Technology*, 1, (2010), 397-414. https://doi.org/10.1146/annurev.food.102308.124134.

European Food Safety Authority. "Quantitive risk assessment of Listeria monocytogenes in a traditional RTE product. *EFSA Journal*, *17*, (2019), e170906, https://doi.org/10.2903/j.efsa.2019.e17090.

Even, Sergine, Lindley, Nic D., Loubière, Pascal, Muriel Cocaign-Bousquet. "Dynamic response of catabolic pathways to autoacidification in *Lactococcus lactis*: transcript profiling and stability in relation to metabolic and energetic constraints." *Molecular Microbiology*, 45, (2002), 114-1152. http://dx.doi.org/10.1046/j.1365-2958.2002.03086.x.

Gharsallaoui, Adem, Oulahal, Nadia, Joly, Catherine, Degraeve, Pascal. "Nisin as a food preservative: Part 1: Physicochemical properties, antimicrobial activity, and main uses. *Critical Reviews in Food Science and Nutrition*, 56, no 8 (2016), 1262–1274. https://doi.org/10.1080/10408398.2013.763765.

Harzallah, Daoud, Hani Belhadj. "Lactic Acid Bacteria as Probiotics: characteristics, selection criteria and role in immunomodulation of human GI muccosal barrier." In *Lactic Acid Bacteria - R & D for Food, Health and Livestock Purposes*, J. Marcelino Kongo, 197-216. IntechOpen, 2013.

Héchard, Yann, Dérijard, Benoit, Letellier, François, Yves Cenatiempo. "Characterization and purification of mesentericin Y105, an anti-*Listeria* bacteriocin from *Leuconostoc mesenteroides*." *Journal of General Microbiology*, 138, no. 12, (1992), 2725-2731. https://doi.org/10.1099/00221287-138-12-2725.

Hwang, In-Chan, Oh, Ju K; Kim, Sang H; Oh, Sejong, Dae-Kyung Kang. "Isolation and characterization of an anti-listerial bacteriocin from *Leuconostoc lactis* SD501." *Korean Journal for Food Science of Animal Resources*, 35, no. 5, (2018), 1008-1018. https://doi.org/10.5851/kosfa.2018.e33.

Jordan, Kieran; Olivia McAuliffe. *Listeria monocytogenes* in Foods. *Advances in Food and Nutrition Research*, 86, (2018), 181-213. https://doi.org/10.1016/bs.afnr.2018.02.006.

Lasik‑Kurdyś, Małgorzata, Anna Sip. "Evaluation of the antimicrobial activity of bacteriocin-like inhibitory substances of enological importance produced by *Oenococcus oeni* isolated from wine." *European Food Research and Technology*, 245, no. 2, (2019), 375-382. https://doi.org/10.1007/s00217-018-3194-1.

Liu, SQ. "Lactic acid bacteria: *Leuconostoc* spp." *Reference Module in Food Science*, (2016). https://doi.org/10.1016/B978-0-08-100596-5.00859-3.

Marco, Maria L., Heeney, Dustin, Binda, Sylvie, Cifelli, Christopher J., Cotter, Paul D, Foligné, Benoit, Gänzle, Michael, Kort, Remco, Pasin, Gonca, Pihlanto, Anne, Smid, Eddy J; Robert Hutkins. "Health benefits of fermented foods: microbiota and beyond." *Current Opinion in Biotechnology*, 44, (2017), 94-102. https://doi.org/10.1016/j.copbio. 2016.11.010.

Papadimitriou, Konstantinos, Alegría, Ángel, Bron, Peter A., de Angelis, Maria, Gobbetti, Marco, Kleerebezem, Michiel, Lemos, José A., Linares, Daniel M., Ross, Paul, Stanton, Catherine, Turroni, Francesca, van Sinderen, Douwe, Varmanen, Pekka, Ventura, Marco, Zúñiga, Manuel, Tsakalidou, Effie, Jan Kok. "Stress physiology of lactic acid bacteria." *Microbiology and Molecular Biology Reviews*, 80, no. 3, (2016), 837–890. https://doi.org/10.1128/MMBR.00076-15.

Papagianni, Maria, Emmanuel M. Papamichael. "Purification, amino acid sequence and characterization of the class IIa bacteriocin weissellin A, produced by *Weissella paramesenteroides* DX." *Bioresource Technology*, 102, no. 12, (2011), 6730-6734. https://doi.org/10.1016/ j.biortech.2011.03.106.

Peng, Chanthol, Borges, Sandra, Magalhães, Rui, Carvalheira, Ana, Ferreira, Vânia, Casquete, Rocío, Paula Teixeira. "Characterization of anti-listerial bacteriocin produced by lactic bacteria isolated from traditional fermented foods from Cambodia." *International Food Research Journal*, 24, no 1, (2017), 386-393.

Ray, Ramesh C; Vinod Joshi. "Fermented Foods: Past, Present and Future." In *Microorganisms and Fermentation of Traditional Foods*, edited by Ramesh C. Ray and Montet Didier, 1-36. CRC Press, 2014.

Rezac, Shannon, Kok, Car Reen, Heermann, Melanie, Robert Hutkins. "Fermented Foods as a Dietary Source of Live Organisms." *Frontiers in Microbiology*, 9, (2018), 1785. https://doi.org/10.3389/ fmicb.2018. 01785.

Sambrook, Joseph, Fritsch, Edward F; Tom Maniatis. "Molecular cloning: A Laboratory Manual (second edition)." In *Journal of Basic Microbiology*, edited by Horst Malke. New York, Cold Spring Harbor Laboratory Press, 1989.

Schägger, Hermann, Gebhard Von Jagow, G. "Tricine-sodium dodecyl sulphate polyacrylamide gel electrophoresis for the separation of protein in the range from 1 to 100 kDa." *Analytical Biochemistry*, 166 no. 2, (1987), 368-379.

Stiles, Michael E. "Bacteriocins produced by *Leuconostoc* species." *Journal of Dairy Science*, 77, no. 9, (1994), 2718–2724. https://doi.org/10.3168/jds.S0022-0302(94)77214-3.

Tomé, Elisabetta, Teixeira, Paula, Paul A. Gibbs. "Anti-listerial inhibitory lactic acid bacteria isolated from commercial cold smoked salmon". *Food Microbiology*, 23, no. 4, (2006), 399-405. https://doi.org/10.1016/j.fm.2005.05.004.

Van Reenen, Carol A; Dicks, Leon MT; Michael L Chikindas. "Isolation, purification and partial characterization of plantaricin 423, a bacteriocin produced by *Lactobacillus plantarum*." *Journal of Applied Microbiology*, 84, no. 6, (1998), 1131-1137. https://doi.org/10.1046/j.1365-2672.1998.00451.x.

World Health Organization (WHO). "The burden of foodborne diseases in the WHO European Region.", 2017, accessed June 1, 2020, https://www.euro.who.int/__data/assets/pdf_file/0005/402989/50607-WHO-Food-Safety-publicationV4_Web.pdf.

Yang, Rongguang, Johnson, Monty, Bibek Ray. "Novel method to extract large amounts of bacteriocins from lactic acid bacteria." *Applied and Environmental Microbiology*, 58, no. 10, (1992), 3355-3359.

Yang, Shih-Chun, Lin, Chih-Hung, Sung, Calvin T; Jia-You Fang. "Antibacterial activities of bacteriocins: application in foods and pharmaceuticals." *Frontiers in Microbiology*, 5, (2014), 241. https://doi.org/10.3389/fmicb.2014.00241.

In: Fermented Foods
Editor: Oliver Kovalyov
ISBN: 978-1-53618-656-7

Chapter 4

FERMENTED MEATS: MICROBIAL DISTRIBUTION, NUTRITIONAL VALUES AND HEALTH ASPECTS

Sujitra Techo[1,*], Engkarat Kingkaew[2] and Somboon Tanasupawat[2,†]

[1]Mahidol University, Nakhonsawan Campus, Nakhonsawan, Thailand
[2]Department of Biochemistry and Microbiology, Faculty of Pharmaceutical Sciences, Chulalongkorn University, Bangkok, Thailand

ABSTRACT

Fermented meats are traditional and unique foods which have long been known for centuries. A diverse kind of fermented meat products with different characteristics are produced using widely different formulations, and manufacturing processes involving the nature and tradition of the countries. The microorganisms play a significant role on

---

* Corresponding Author's Email: sujitra.tac@mahidol.ac.th.
† Corresponding Author's Email: Somboon.T@chula.ac.th.

fermentation process to provide the desirable quality attributes of products. Two major groups of microorganisms participated in meat fermentation include lactic acid bacteria (LAB) and coagulase negative staphylococci (CNS). Moreover, yeasts and molds also involve in the fermentation. Fermented meats are rich sources of protein and fat. They are also the important sources of several other micronutrients such as vitamins (niacin, pyridoxin, thiamine, and riboflavin) and trace elements (zinc, iron, and selenium). However, fermented meats consumption may provide a negative effect on health as a consequence of high levels of fat, cholesterol and low polyunsaturated/saturated fatty acid ratio (PUFA/SFA). Substitution of meat fat with plant-based fat is used to reduce the risk of health disease. Several techniques including addition of dietary fiber, probiotics, synbiotics, acid whey and functional starter cultures are applied to enhance the nutritional values of these products. Fermented meats exhibit several bioactive activities such as antioxidant activity, angiotensin converting enzyme (ACE) inhibitory activity, antidiabetic activity and low cytotoxicity. Therefore, fermented meats are considered as a kind of functional food that can provide a positive effect on human health, particularly when probiotics are used in the manufacturing process. Consumers are aware of safety concerns of fermented meats due to the risk of microbiological, namely foodborne pathogens and chemical hazards especially, biogenic amines, nitrosamines, mycotoxins, and polycyclic aromatic hydrocarbons (PAHs). Starter cultures are employed to ensure the safety of fermented meat products by inhibition of the spoilage and/or pathogenic microorganisms and reduction of the chemical additives particularly, nitrates and nitrites.

Keywords: fermented meats, bioactive activities, lactic acid bacteria, staphylococci

1. Introduction

Fermented meats have spread from the Mediterranean to Northern Europe from where migration established production in the United States, South America, Australia, and elsewhere (Adams, 1986). Meat comprises of water, protein, lipid, minerals, and trace amounts of carbohydrate (Toldrá and Reig, 2015). Fermented meats are highly nutritious and important source of protein. They contain essential amino acids which may absent in other protein sources such as grains and legumes (Adewumi,

2019). The meat products have been consumed for centuries in many countries and they are important food sources (Liu et al., 2011; Lucke, 1994). Fermented meats are made from comminuted meat and fat, mixed with salt, seasonings, spices, and authorized additives that are then stuffed into casings, acidified through lactic-acid fermentation, ripened, dried, and sometimes smoked. The products are classified based on acidity, degree of mincing, use and type of starter cultures, spices and seasonings, diameter and type of casing, degree of drying and smoking. The addition of curing agents (nitrate and/or nitrite) to provide the characteristic red color and the action of microorganisms to produce lactic acid is a common trait. During fermentation and ripening, various complex physical and chemical reactions determine the characteristic tangy flavor, texture, cured meat color, and increase shelf-life of this type of sausage (Campbell-Platt and Cook, 1995; Toldrá, 2007).

The fermetation of products, Italian salami in Italy (Pisacane et al., 2015), Slavonski kulen in Croatia (Babić et al., 2011), fermented llama sausages in Argentina (Fontana et al., 2016), Chorizo de Léon (Spanish dry fermented sausage) in Spain (Quijada et al., 2018), fermented Swiss meat products in Switzerland (Marty et al., 2012), Chinese dry-cured sausage and smoked-cured sausage in China (Wang et al., 2018), are involved with different types of microorganisms such as lactic acid bacteria and staphylococci while Nham (Santiyanont et al., 2019), Mum (Wanangkarn et al., 2014) in Thailand, Nem chua in Vietnam are reported to involve with lactic acid bacteria. During fermentation, the quality and safety of products are improved by the acid production (Lucke, 1998). In addition, the different biochemical potential lead to modify the organoleptic profile and enhance nutritional value, keeping quality as well as stability of end product in meat fermentation (Leroy et al., 2006). The production of fermented meats has been increasing rapidly and the meat industry is looking for functional starter cultures with improved sensory attributes, nutritional quality, and health and microbial safety of products. In addition, the incorporation of probiotic strains in fermented meats has been established for the development of healthy products.

2. Microbial Distribution

Lactic acid bacteria (LAB) and coagulase-negative staphylococci (CNS) are the major groups of bacteria involving in the manufacturing of fermented meats. They provide the final products with more hygenic and sensory qualities. LAB produce lactic acid through the hexose fermentation consequently, the pH of product is decreased. CNS exerts lipolysis, proteolysis and free amino acid decomposition that contribute to the development of color and flavor of fermented meats (Talon et al., 2007). In order to select potential autochthonous functional starter cultures, LAB and CNS microbiota that play a vital role in fermentation are widely investigated.

2.1. Lactic Acid Bacteria (LAB)

Studies on LAB microbiota presented in fermented meat products are generally based on the molecular methods as shown in Table 1. The most abundant LAB found in fermented meats belong to the genus *Lactobacillus* (*Lb.*). The others including *Enterococcus* (*En.*), *Leuconostoc* (*Leuc.*), *Lactococcus* (*Lc.*), *Weissella* (*W.*), *Pediococcus* (*Ped.*), *Streptococcus* (*Strep.*), *Carnobacterium* (*C.*) and *Vagococcus* (*V.*) are also detected. According to the LAB community studies, the most commonly LAB identified in traditional fermented meats was *Lb. sakei*. Charmpi et al., (2020) reported that *Lb. sakei* (57%), *Carnobacterium* sp. (29%), and *Lb. curvatus* (14%) were presented at the initial stage of the fermentation of pork of normal raw meat quality. *Lb. sakei* was the most predominant throughout the fermentation process. An insignificant number of *Lc. lactis* was also found at 0%, 1%, and 2% of salt levels. LAB strains identified by API test and RAPD-PCR (random amplification of polymorphic DNA-polymerase chain reaction) in *Slavonski kulen*, a traditional Croatian fermented sausage, were *Lb. delbrueckii*, *Lb. acidophilus*, *Lb. plantarum* and *Leuc. mesenteroides*. *Leuc. mesenteroides* and *Lb. acidophilus* (Babić et al., 2011). All LAB isolates (917 isolates) collected from spontaneously fermented wild boar and deer meat sausages produced in Croatia were genotyped by rep-PCR (repetitive-PCR) and identified by 16S rRNA gene sequencing. *Leuc. mesenteroides* was identified as the most prominent species (259 isolates), followed by *Lb. sakei* (190 isolates) and *En. casseliflavus* (106 isolates) (Maksimovic et al., 2018). After molecular identification of 219 lactobacilli isolated from traditional Italian salami, *Lb. sakei* and *Lb. curvatus* were the most prevalent species. *Lb. sakei* coming from casing was appeared in the salami at different times of ripening (Pisacane et al., 2015). *En. faecium*, *Lb. sakei*, *Lc. lactis*, *Lactococcus* sp. were identified by RT-PCR-DGGE during the manufacturing and ripening of Fabriano-like sausage prepared without the addition of starter cultures (Cardinali et al., 2018). Two hundred and sixty-three LAB were isolated and identified by PCR-DGGE methods and 16S rRNA gene sequencing. *Lb. sakei* was the dominant species found in *Pitina* (81.8% of total isolates). *C. divergens*, *Strep. equinus*, *Leuc. mesenteroides* and *Lc. lactis* were also identified at the levels of 10.6%, 2.7%, 2.3% and 2%, respectively (Iacumin et al., 2020). Santos et al. (2017) have investigated the enterococci group which are known to the native microbiota in traditional dry smoked fermented sausages of Portugal. *Catalão*, *Chouriço-preto* and *Linguiça* samples used in the study contained the number of enterococci at 10^4 cfu/g. *Salsichão* and *Paio* carried low number of enterococci at 10^2 cfu/g. From a total of 147 isolates, 73 were identified as *En. faecalis*, 60 as *En. faecium* and 14 as *En. durans*. Five different *Lactobacillus* species were found during the processing of *Chorizo de León*. *Lb. sakei* (93.0%) was the main species, followed by *Lb. fuchuensis* (2.3%), *Lb. plantarum* (2.2%), *Lb. curvatus* (2.0%) and *Lb. algidus* (0.5%) (Quijada et al., 2018). *Asturian Chorizo* is a traditional fermented sausage from Northern Spain, produced in a traditional approach, without starters. Based on genotypic characterization by 16S rRNA sequencing, 75 isolates of LAB isolated from *Asturian Chorizo* were identified. All isolates belonged to the genus *Lactobacillus*. the most common species were *Lb. plantarum* (51%), *Lb. sakei* (36%), *Lb. futsai* (5%), *Lb. curvatus* (3%) and *Lb. graminis* (3%) (Prado et al., 2019). LAB associated during fermention process of fermented Swiss meat products manufactured with game meat are *Lb. sakei*, *Lb. curvatus*, *En. faecalis*, *Enterococcus* sp. and *Streptococcus* sp. In addition to species mentioned above, *En. faecium and Pediococcus* sp. is detected in fermented end products by using PCR-RFLP technique (Marty et al., 2012). A high throughput sequencing approach conducted by Illumina MiSeq™ analysis was used to assess the bacterial communities in Chinese dry-cured sausage and Chinese smoked-cured sausage. Lactic acid bacteria (LAB) including *Lactobacillus* spp., *Weissella* spp., *Pediococcus* spp., and *Lactococcus* spp. were the main bacteria in both products. *Enterococcus* spp. was the minor genus found in these products (Wang et al., 2018). The study of Lv et al., (2019) have investigated microbial diversity of sour meat or a traditional fermented meat product of China by using MALDI-TOF MS. *Lactobacillus* was the major genus found in sour meats prepared at different temperature. *Lb. sakei* was the predominant species at the early stage of fermentation whereas *Lb. curvatus* appeared to the main species at the middle stage and *Lb. plantarum* became to the dominant species at the later stage. A total of 630 LAB were isolated from *Mum* (Thai fermented sausages) during fermentation and storage. The prominent species involved in *Mum* was *Lb. sakei* accounting as 30% from all isolates, followed by *Lb. plantarum* (24.76%), *Leuc. mesenteroides* (13.33%), *Lb. brevis* (11.90%), *Ped. pentosaceus* (7.62%), *Lc. lactis* (7.62%), and *Lb. fermentum* (4.76%) (Wanangkarn et al., 2014). PCA and MRS media were used to isolate the LAB from Thai fermented pork sausages (*Nham*). *Lc. garvieae*, *Lb. plantarum, Ped. pentosaceus*, *W. cibaria*, *W. confusa*, *W. hellenica*, *W. paramesenteroides*, *Lb. brevis*, *Lb. sakei*, *Lb. formosensis/farciminis*, *Lc. lactis*, *Leuc. citreum*, *Leuc. fallax*, *Strep. parauberis*, *V. teuberi* are the species associated with *Nham* fermentation. The most dominant species identified by the culture-dependent method were *Ped. pentosaceus* on day 3 and *Lb. plantarum* during days 5-10 while *Weissella* was the third most abundant genus. The results from culture-dependent method are similar to those received from pyrosequencing. Four major LAB genera such as *Weissella*, *Lactococcus*, *Lactobacillus*, and *Pediococcus* were dominated during fermentation (Santiyanont et al., 2019). *Lb. pentosus*, *Lb.*

plantarum, *Lb. brevis*, *Lb. paracasei*, *Lb. fermentum*, *Lb. acidipiscis*, *Lb. farciminis*, *Lb. rossiae*, *Lb. fuchuensis*, *Lb. namurensis*, *Lc. lactis*, *Leuc. citreum*, *Leuc. fallax*, *Ped. acidilactici*, *Ped. pentosaceus*, *Ped. stilesii*, *W. cibaria* and *W. paramesenteroides*, *Lb. helveticus/crispatus*, *Lc. garvieae* and *Vagococcus* sp. are involved during fermentation of *Nem chua* (a Vietnamese fermented meat product). They are identified by both culture-dependent and culture-independent methods (Nguyen et al., 2013). *Lb. algidus*, *Lb. alimentarius*, *Lb. crustorum*, *Lb. fuchuensis*, *Lb. sakei*, *Leuc. carnosum*, *Leuc. citreum*, *Leuc. gasicomitatum*, *Leuc. inhae*, *Leuc. kimchii*, *Leuc. lactis*, *Leuc. mesenteroides*, *Strep. viridans*, *Strep. infantis*, *Leuc. garvieae*, *Lc. piscium*, *En. casetiflavus*, *W. ghanensis*, *W. viridescens* strains were involved in the ripening of naturally fermented llama meat sausages. *Lb. sakei* was largely dominated in both fermentation process of pilot-plant scale and an artisanal one (Fontana et al., 2016).

2.2. Coagulase-Negative Staphylococci (CNS)

Coagulase-negative staphylococci (CNS) are widely distributed in fermented meats. The most abundant species seems to be *S. equorum*, followed by *S. xylosus* as presented in Table 1. *S. saprophyticus* (61%), *S. equorum* (28%), and *S. xylosus* (11%) were remarkable detected in all batches of meat fermentation. The species *S. epidermidis*, *S. pasteuri* were also encountered, and *S. aureus* emerges sporadically in the batches with salt levels of 2% and 4% (Charmpi et al., 2020). According to biochemical (API) tests and RAPD-PCR analysis, *S. xylosus*, *S. auricularis*, *S. warneri* and *S. lentus* were the species observed in the production of *Slavonski kulen* (Babić et al., 2011). Several species of *Staphylococcus* were isolated and identified during the maturation process of traditional Italian salami. *S. xylosus* and *S. saprophyticus* were the most frequently isolated (Pisacane et al., 2015). *S. equorum* was detected at the 42-days during ripening of Fabriano-like sausage processed without starter cultures (Cardinali et al., 2018). The prominent species detected in *Pitina* belonged to *S. xylosus* (143 isolates). The other species included *S. equorum* (69 isolates), *S.*

succinus (38 isolates) and *S. warneri* (17 isolates). A low percentage (≤ 1%) of isolates was found to be *S. saprophyticus*, *S. haemolyticus*, *S. hominis* and *S. epidermidis* (Iacumin et al., 2020). There are five different *Staphylococcus* species observed in *Chorizo de León* (a high-value Spanish dry fermented sausage). The highest percentage of *S. xylosus* (84.5%) was found, followed by by *S. sciuri* (8.5%), *S. saprophyticus* (3.2%), *S. equorum* (1.9%) and *S. condimenti* (1.9%) (Quijada et al., 2018). *Staphylococcus* 236 isolates from 21 samples of naturally fermented Swiss meat products were subjected to identified by using PCR-RFLP. *S. equorum* was the dominant species (20.3%) followed by *S. warneri* (16.5%), *S. saprophyticus* (15.3%), *S. epidermidis* (13.1%) and *S. xylosus* (11.4%). Each of the remaining ten species contributed less than 7% to the staphylococcal microbiota. The percentage of the pathogenic *S. aureus* was 6.8% of total staphylococci (Marty et al., 2012). Wang et al., (2018) reported that the highest number of operational taxonomic units (OTUs) for *Staphylococcus* spp., was found in Chinese dry-cured sausage and Chinese smoked-cured sausage for 97.45%. Santiyanont et al. (2019) reported that *S. cohnii* was found at day 0 of fermentation of *Nham* but afterward this species was disappeared throughout the fermentation period.

3. Nutritional Values

Meat is the major ingredient in fermented meat manufacturing. Meat comprises of water (72-76%), protein (15-22%), lipid (1.5-4%), minerals (<1%), and trace amounts of carbohydrate (Toldrá and Reig, 2015). Hence, fermented meats are highly nutritious and important source of protein. They contain essential amino acids which may absent in other protein sources such as grains and legumes (Adewumi, 2019). As a result of muscle and microbial peptidases during fermentation and ripening period, proteolysis is generated and play an important role in the degradation of proteins (Gallego et al., 2018). Proteolysis contributes to the generation of

Table 1. LAB and CNS associated in fermented meat products

Country	Product	LAB	CNS	References
Belgium	Fermented meat	*Lb. sakei*, *Carnobacterium* sp., *Lb. curvatus*, *Lc. lactis*	*S. saprophyticus*, *S. equorum*, *S. xylosus*, *S. saprophyticus*, *S. epidermidis*, *S. pasteuri*, *S. aureus*	Charmpi et al. (2020)
Croatia	*Slavonski kulen*	*Lb. delbrueckii*, *Lb. acidophilus*, *Lb. plantarum*, *Leuc. mesenteroides*	*S. xylosus*, *S. warneri*, *S. lentus*, *S. auricularis.*	Babić et al. (2011)
	Fermented wild boar and deer meat sausages	*Leuc. mesenteroides*, *En. casseliflavus*, *Enterococcus* spp., *En. faecalis*, *W. viridescens*, *Lb. sakei*, *En. gallinarum*, *En. faecium*, *Streptococcus* spp., *Lc. garviae*, *Weissella* spp., *En. durans*, *Strep. gallolyticus*, *Lactobacillus* spp., *En. hirae*, *Leuconostoc* spp., *Lb. curvatus*, *C. maltaromaticum*, *C. divergens*, *P. pentosaceus*, *W. helenica*	ND	Maksimovic et al. (2018)
Italy	Traditional Italian salami	*Lb. sakei*, *Lb. plantarum*, *Lb. salivarius*, *Lb. paracasei*, *Lb. curvatus*, *Lb. fermentum*	*S. xylosus*, *S. saprophyticus*, *S. pasteuri*, *S. epidermidis*, *S. equorum*, *S. simulans*	Pisacane et al. (2015)
	Fabriano-like fermented sausages	*En. faecium*, *Lb. sakei*, *Lc. lactis*, *Lactococcus* sp.	*S. arlettae*/*S. saprophyticus*, *S. equorum*	Cardinali et al. (2018)
	Pitina (fermented sausages)	*Lb. casei*, *En. faecalis*, *Lc. lactis*, *Leuc. mesenteroides*, *Strep. equinus*, *C. divergens*, *Lb. sakei*	*S. epidermis*, *S. hominis*, *S. haemolyticus*, *S. saprophyticus*, *S. warneri*, *S. succinus*, *S. equorum*, *S. xylosus*	Iacumin et al. (2020)
Portugal	Traditional dry smoked fermented meat sausages	*En. faecalis*, *En. faecium*, *En. durans*	ND	Santos et al. (2017)

Table 1. (Continued)

Country	Product	LAB	CNS	References
Spain	Chorizo de Léon (Spanish dry fermented sausage)	*Lb. sakei*, *Lb. fuchuensis*, *Lb. plantarum*, *Lb. curvatus*, *Lb. algidus*	*S. xylosus*, *S. sciuri*, *S. saprophyticus*, *S. equorum*, *S. condimenti*	Quijada et al. (2018)
	Asturian Chorizo	*Lb. plantarum, Lb. sakei, Lb. futsai, Lb. curvatus, Lb. graminis*	ND	Prado et al. (2019)
Switzerland	Fermented Swiss meat products	*Lb. sakei*, *Lb. curvatus*, *En. faecalis*, *Enterococcus* sp., *Streptococcus* sp.	*S. aureus*, *S. saprophyticus*, *S. epidermidis*, *S. xylosus*, *S. equorum*, *S. vitulinus*, *S. succinus*, *S. warneri*, *S. capitis*, *S. sciuri*, *S. simulans*, *S. gallinarum*, *S. chromogenes*	Marty et al. (2012)
China	Chinese dry-cured sausage and smoked-cured sausage	*Lactobacillus* spp., *Weissella* spp., *Pediococcus* spp., *Lactococcus* spp., *Enterococcus* spp.	*Staphylococcus* spp.	Wang et al. (2018)
	Sour meats	*Lb. sakei*, *Lb. curvatus*, *Lb. plantarum*	*S. equorum*, *Staphylococcus* spp.	Lv et al. (2019)
Thailand	*Mum*	*Lb. plantarum*, *Lb. sakei*, *Lb. fermentum*, *Lb. brevis*, *Ped. pentosaceus*, *Leuc. mesenteroides*, *Lc. lactis*	ND	Wanangkarn et al. (2014)
		Ped. pentosaceus	ND	Tanasupawat et al. (2015)
	Nham	*Lc. garvieae*, *Lb. plantarum, Ped. pentosaceus*, *W. cibaria*, *W. confusa*, *W. hellenica*, *W. paramesenteroides*, *Lb. brevis*, *Lb. sakei*, *Lb. formosensis/farciminis*, *Lc. lactis*, *Leuc. citreum*, *Leuc. fallax*, *Strep. parauberis*, *V. teuberi*	*S. cohnii*	Santiyanont et al. (2019)

Country	Product	LAB	CNS	References
Thailand	*Nham*	*Lb. pentosus, W. thailandensis,* and *Ped. pentosaceus*	ND	Tanasupawat et al. (2015)
	Sai-krog-prieo	*Ped. pentosaceus, Leuc. fallax, Lb. pentosus, Ped. lolii, W. cibaria,* and *W. paramesenteroides*	ND	Tanasupawat et al. (2015)
		W. confusa, *W. kimchii*, *Ped. pentosaceus*, *Ped. acidilactici*, *Lb. fermentum*, *Lb. brevis*, *Lb. farciminis*, *Lb. plantarum* and *Lb. sakei*	ND	Phalakornkule and Tanasupawat (2006)
		En. durans, *Lb. plantarum*, *Lb. brevis* and *Ped. pentosaceus*	ND	Wanangkarn et al. (2020)
Vietnam	*Nem chua*	*Lb. pentosus*, *Lb. plantarum*, *Lb. brevis*, *Lb. paracasei*, *Lb. fermentum*, *Lb. acidipiscis*, *Lb. farciminis*, *Lb. rossiae*, *Lb. fuchuensis*, *Lb. namurensis*, *Lc. lactis*, *Leuc. citreum*, *Leuc. fallax*, *Ped. acidilactici*, *Ped. pentosaceus*, *Ped. stilesii*, *W. cibaria*, *W. paramesenteroides*, *Lb. helveticus/crispatus*, *Lc. garvieae*, *Vagococcus* sp.	ND	Nguyen et al. (2013)
Argentina	Fermented llama sausages	*Lb. algidus*, *Lb. Alimentarius*, *Lb. crustorum*, *Lb. fuchuensis*, *Lb. sakei*, *Leuc. carnosum*, *Leuc. citreum*, *Leuc. gasicomitatum*, *Leuc. inhae*, *Leuc. kimchii*, *Leuc. lactis*, *Leuc. mesenteroides*, *Strep. viridans*, *Strep. infantis*, *Leuc. garvieae*, *Lc. piscium*, *En. casellflavus*, *W. ghanensis*, *W. viridescens*	ND	Fontana et al. (2016)

ND; not determined.

large amounts of peptides and free amino acids (FAAs). These compounds are related not only to sensory characteristics of the final product but also to other properties and functionalities such as biological activities (Toldrá et al., 2018). Proteolysis of both myofibrillar and sarcoplasmic proteins in *Nham*, a Thai-style fermented pork sausage, occurred as a result of fermentation. At the beginning of hydrolysis, endogenous cathepsins mainly performed to hydrolyze muscle protein and followed by the action of bacterial enzymes, which actively degraded oligopeptides to small peptides and FAAs. Moreover, TCA-soluble peptides and FAAs increased during *Nham* fermentation which associated with the development of flavor and aroma in *Nham* (Visessanguan et al., 2004). There are differences in the degree of proteolysis generating during the ripening of three European dry-fermented sausages produced in Spain, Italy and Belgium. This is due to different variables such as product formulation, processing conditions and microbial population. Total concentrations of FAAs were 1854, 2235, and 2414 mg/100 g dry matter, in Spanish, Italian and Belgian dry-fermented sausages, respectively. Free amino acid profiles were found to be different among three-types of European dry-fermented sausages. These can be explained that the activity of enzymes, either of muscle and bacterial origin, is influenced by processing parameters such as the temperature, relative humidity, length of fermentation and ripening times, changes in pH, content of salt, nitrate and nitrite, and type and content of carbohydrates (Gallego et al., 2018). The resulting peptides and amino acids have widely known as bioactive substances which exhibited the physiological functions in human body such as antioxidant, antihypertensive, antimicrobial, anticancer, antithrombotic, anti-diabetic, immunomodulating and probiotic (Pogorzelska-Nowicka et al., 2018)

Fat is the essential constituent for fermented meats. It provides energy, satisfactory quality and sensory characteristics especially, chewiness, juiciness and aroma of products. Fermented meats produce from different kinds of meat including beef, pork, chicken, camel, foal or even game meat which depend on the formulations, geographical regions and/or tradition of the countries. Differences in fat level of 4 types of Argentinian beef mixed with pork sausages with different composition and the processing according to the manufacturing area have been reported. *Salamín* (dry cured sausage), *Chorizo* (raw sausage), *Morcilla* (cooked sausage) and *Chorizo ahumado* (smoked sausage) contain 29.71%, 17.70%, 17.65% and 18.50% of total fat (Romero et al., 2013). Types of meat and fat used in the recipes had an effect on chemical composition of Turkish dry-fermented sausage (*Sucuk*). Fat content in four *Sucuk* samples varied from 31.39% to 39.28%. *Sucuk* reformulated with camel meat and hump had the highest content of fat (Kargozari et al., 2014). However, high animal fat content in fermented meat products may has a negative impact on human health. Moreover, high amount of cholesterol and low polyunsaturated/saturated fatty acid ratio (PUFA/SFA) are the causes of coronary diseases (Muguerza et al., 2001). For decades, consumers have interested in healthier meat products. To meet with consumer's need, many researches have attempted to reduce fat level and decrease in saturated fatty acids (SFAs), with a concomitant increase in polyunsaturated fatty acids (PUFAs) or monounsaturated fatty acids (MUFAs). Different types of thermal processing including microwave, skillet, oven, charcoal grill, electric fryer and electric grill effected on fat and cholesterol content of *Alheiras* (Portugese fermented meat sausage). Charcoal grill was the best approach to obtain *Alheiras* with less fat (12.5 g/100 g) and cholesterol (29.3 mg/100 g), corresponding to a lower caloric intake (231.8 kcal, less 13% than the raw ones). Inversely, fried *Alheiras* presented the worst nutritional profile, with the highest levels of fat (18.1 g/100 g) and cholesterol (76.0 g/100 g) (Campos et al., 2013). The replacement of pork backfat with two oleogels, a mixture of γ-oryzanol and β-sitosterol (SO) and beeswax (B), was performed in dry-cured sausages. The fatty acid profile improved in terms of nutrition: the PUFA/SFA and n-6/n-3 ratio was about 1.41 and 0.93 for the higher levels of replacement at 40% SO and 40% B, respectively (Franco et al., 2019). Fiber have been added to fermented meat products as fat-replacement. Menegas et al. (2016) reported that adding inulin (7%) as partial substitute for corn oil in the processing of dry-fermented chicken sausages resulted in the reduction 52.9% of lipid content and increased nutritional value (higher amount of soluble fibers and protein) compared to the product with standard oil content. In order to reduce fat content and/or to improve the fatty acid profile, partial replacement of pork backfat in dry-fermented sausages by oil mixture (olive, linseed and fish oil) stabilized in a konjac matrix has been studied. Low-fat sausages contained 99-130 g/kg of fat versus 316 g/kg in the normal fat sausages. Addition of an oil-in-konjac matrix reduced SFA content and increased PUFA, improving the n-6/n-3 ratio in dry sausage (Jiménez-Colmenero et al., 2013). Substitution of pork back-fat by pure or commercial encapsulated fish oil (15% and 30%) had the effect on fat composition of Dutch-style fermented sausages. Modified sausages had a lower SFA level approximately 3–8% compared to the controls. PUFA increased 2–8% compared to controls the ratio of n-6/n-3 fatty acids decreased by a factor of 3 to 10 compared to the control sausage, depending on the formulation. The lower this ratio, the better the product is in terms of health (Josquin et al., 2012).

4. Health-Promoting Activities

Over the last decades, consumers prefer the food product exerted high nutritional values and health-promoting activities. The incorporation of probiotic strains in fermented meats has been established as a useful tool for the development of healthy products. Proteolysis or protein degradation generally takes place during the manufacturing of fermented sausages. The muscle and microbial peptidases play a major role on protein degradation, generation large amounts of peptides and free amino acids (Gallego et al., 2017). These substances are not only responsible for the flavor development of fermented meats but some peptides can also exhibit bioactive properties such as antioxidant, ACE inhibitory, antidiabetic activities and cytotoxicity.

4.1. Probiotics

Nowadays, the meat industry has focused on the development of meat products enhancing quality and functionality as a consequence of interesting of functional foods by consumers. Probiotic food products are the health-promoting foods exerting health benefits to the host by the action of selected LAB when administered in adequate amounts (10^6 – 10^7 cfu/ml) (FAO/WHO, 2002). Due to the fact that fermented meats are regularly consumed in uncooked form, they are considered to be the functional foods that deliver lived microorganisms, particularly LAB. Several studies have attempted to screen a candidate probiotic for further use as starter culture. In order to select a candidate probiotic culture, the safety of strains, the probiotic properties and technological characteristics have to be assessed. Safety considerations include the origin, non-pathogenicity and antibiotic resistance pattern. Functional aspects include viability and persistence in the GI-tract, immunomodulation, antagonistic and antimutagenic properties. Probiotics must be able to be manufactured under industrial conditions. Moreover, they have to survive and retain their functionality during storage, and also in the foods into which they are incorporated without producing off-flavours (Saarela et al., 2000).

Most strains applied as probiotic cultures in fermented meats belonged to genus *Lactobacillus* as shown in Table 2. The others included the genera *Pediococcus*, *Weisella* and *Bifidobacterium* (*B.*). Among 22 strains isolated from 15 different Scandinavian-type fermented sausages, only three strains, *Lb. plantarum* MF1291, MF1298, *Lb. pentosus* MF1300, had a potential as probiotics according to their survival following exposure to acid (HCl, pH 2.5) and oxgall (0.3%, w/v) for 4 h. They also adhered to the human colon adenocarcinoma cell line Caco-2 and expressed the antimicrobial activity against potential pathogens. Moreover, high viability (4.7×10^7 - 2.9×10^8 cfu/g) were observed throughout the end of processing. The three strains were all successfully applied as starter cultures for the production of fermented sausages (Klingberg et al., 2005). Ruiz-Moyano et al. (2010) reported that *Ped. acidilactici* SP979 possessed an ability to compete against enteropathogens such as *Salmonella choleraesuis* on the intestinal epithelium and an inhibition of such pathogenic bacteria as *Listeria monocytogenes* and *Staphylococcus aureus in vitro*. This strain was also considered as safe by its void aminogenic potential, low D-lactic

Table 2. Probiotic cultures used in fermented meat products

Probiotic strain	Product	Results/Findings	References
Lb. plantarum MF1291, MF1298, *Lb. pentosus* MF1300	Scandinavian-type fermented sausages	- Survive upon exposure to pH 2.5 or 0.3% oxgall, adhere to the human colon adenocarcinoma cell line Caco-2 and express the antimicrobial activity	Klingberg et al. (2005)
Ped. acidilactici SP979	Iberian dry fermented sausages	- Competes with enteropathogens such as *Salmonella choleraesuis* and inhibit the growth of *Listeria monocytogenes* and *Staphylococcus aureus in vitro*	Ruiz-Moyano et al. (2010)
W. cibaria N23	*Nham*	-Produces bacteriocin N23 which inhibit the growth of *W. confusa* N31	Pringsulaka et al. (2012)
Lb. brevis SCH6, *Ped. pentosaceus* BAL6 and KL14	Polish raw fermented meat products	-Reveal probiotic characteristics -Be used as viable bioprotective culture in raw fermented meat products as starter cultures	Rzepkowska et al. (2017)
Ped. pentosaceus MZF16	*Dried Ossban*	A probiotic and bioprotective agent for fermented foods, including Tunisian dry meat and sausages	Zommiti et al. (2018)
Ped. pentosaceus LMQS 331.3 and *Ped. acidilactici* LMQS 154.1	Fermented sausages	-Produce bacteriocin, antibacterial potential against *Listeria* spp.	Bungenstock et al. (2020)
Lb. acidophilus CCDM 476 and *B. animalis* 241a.	Mutton fermented sausages	- *Lb. acidophilus* CCDM 476 showed remarkable viability after 60 days - The product has a better texture, and a reduction of the typical smell of mutton	Holko et al. (2013)
Lb. casei CTC1677, *Lb. casei* CTC1678, *Lb. rhamnosus* CTC1679	*Fuet*	- *Lb. rhamnosus* CTC1679 was the only strain able to grow and dominate (levels ca. 10^8 CFU/g) in two independent trials, throughout the ripening process	Rubio et al. (2014a)
Lb. plantarum 299v	Bovine salami	- Shows remarkable viability as probiotic after 60 days storage and has the same level of sensory acceptance of control salami	Blaiotta et al. (2017)
Lb. rhamnosus CTC1679		- Grows, survives and dominates (levels ca. 10^8 CFU/g), prevented the growth of *Listeria monocytogenes* and eliminated *Salmonella*	Rubio et al. (2014b)

Probiotic strain	Product	Results/Findings	References
Lb. acidophilus Bauer, *B. animalis* subsp. *lactis* BB-12, *Lb. rhamnosus* LOCK900	Dry-fermented neck and sausages	-*Lb. acidophilus* Bauer preserves the quality of products better than the *B. animalis* and *Lb. rhamnosus* LOCK900	Wójciak et al., (2017)
Lb. plantarum KACC 92189	Fermented sausages	- Inhibits spoilage bacteria, lowers lipid oxidation level and reduces biogenic amine content	Ba et al., (2018)
Lb. plantarum L125	Fermented sausages	- Leads to equal or better characteristics of products - Survives, competes well with the starter cultures used - Receives similar or higher sensory scores compared to the control (commercial) ones	Pavli et al., (2020)

acid production and antibiotic resistance pattern. *Ped. acidilactici* SP979 can be used individually and in combination with other starter cultures in the manufacturing of Iberian dry-fermented sausages. *W. cibaria* N23 isolated from *nham* showed the production of bacteriocin N23 with antimicrobial activity against *W. confusa* N31. Strain had a potential to use as starter culture in *Nham* production to improve the safety of consumption (Pringsulaka et al., 2012). *Lb. brevis* SCH6, *Ped. pentosaceus* BAL6, and KL14 revealed the resistance to the gastrointestinal conditions, safety assessment, and antimicrobial properties as compared to the other LAB strains. They were viable bioprotective culture that can be inoculated in raw fermented meat products as starter cultures (Rzepkowska et al., 2017). *Ped. pentosaceus* MZF16 had been isolated from *Dried Ossban* showed a high tolerance to gastric stress conditions, and bile salts. Strain also had an ability to adhere to the human intestinal Caco-2/TC7 cell line and resisted to gentamycin and chloramphenicol but did not contain virulence genes. Furthermore, this strain can slightly stimulate the immune system and enhance the intestinal epithelial barrier function. Moreover, this bacterium expressed the anti-*Listeria* activity by the production of bacteriocin (Zommiti et al., 2018). Bungenstock et al. (2020) revealed that the thermostable cell free supernatant of *Ped. pentosaceus* LMQS 331.3 and *Ped. acidilactici* LMQS 154.1 displayed antibacterial effect against

Listeria spp. They were further studied and confirmed for the presence of bacteriocin structural genes. Two strains can be used to enhance product and consumer safety. To obtain safe functional food products with probiotic potential, the suitability of strains as starter culture has also examined in addition to safety and probiotic characteristics. Holko et al. (2013) have attempted to substitute a traditional starter culture in mutton fermented sausages by probiotic strains of *Lb. acidophilus* CCDM 476 and *B. animalis* 241a. Comparison of technological properties of the traditional and the probiotic sausages were investigated. They found that lactobacilli (10^7 cfu/g) and bifidobacteria (10^3 cfu/g) counts in the final product did not affect the technological properties. *Lb. acidophilus* CCDM 476 showed remarkable viability (10^6 cfu/g) of the probiotic strain even after 60 days of storage, whereas the number of *B. animalis* were under detection limit in the long storage period. Moreover, no marked differences detected in the quality of product which were inoculated with *Lb. acidophilus* CCDM 476 and traditional starter cultures. *Lb. rhamnosus* CTC1679 was the most suitable potential probiotic strain for production of reduced Na^+ and fat content low-acid fermented sausages owing to its high viability (10^8 cfu/g) at the end of processing, without affecting the characteristic sensory properties of the product (Rubio et al., 2014a). The addition of *Lb. plantarum* 299v as a probiotic culture in bovine salami production resulted in the same level of sensory acceptance score of probiotic and traditional bovine salami samples. This strain exhibited a marked viability even after 60 days of product storage (Blaiotta et al., 2017). In order to evaluate the viability of *Lb. rhamnosus* CTC1679 to survive through the gastrointestinal tract (GIT), a human intervention study consisting of 5 healthy volunteers who consumed 25 g of probiotic *fuet* a day for 21 days have investigated. The results revealed that the strain can be used as probiotic culture to produce hygienically safe products, showed the high levels (10^8 CFU/g) in ripened fuets and survived the passage through the human GIT during the consumption of the sausages. After 3 days of non-consumption, the strain was still recovered in the faeces of all the volunteers (Rubio et al., 2014b). *Lb. acidophilus* Bauer was suitable strain for dry-fermented pork neck and sausages since the strain prevented the generation of lipid oxidation and discoloration. Hence, this bacterium preserves the better quality of dry-fermented pork neck and sausages when compared with the *B. animalis* and *Lb. rhamnosus* LOCK0900 (Wójciak et al., 2017). The application of *Lb. plantarum* KACC 92189 as a probiotic starter culture for production of fermented sausages can improve product quality and safety by increasing acidification and decreasing the number of spoilage bacteria count. In adiition, this strain does not produce biogenic amines and can significantly improve the eating quality especially the taste and odor of the product when comparing to the commercial starter culture (Ba et al., 2018). Pavli et al. (2020) reported that *Lb. plantarum* L125 can be applied as probiotic starter culture since it provides equal or better technological characteristics of the produced sausages. This strain exerted high survivability during the storage period. The products produced by this strain received similar or higher sensory scores compared to the control.

4.2. Antioxidant Activity

Oxidation is an important reaction in an organism's metabolism. Products resulting from incomplete oxidation of oxygen such as reactive oxygen species (ROS) and other free radicals produce oxidative damage to DNA, proteins, and other macromolecules such as lipids (Nedelcheva et al., 2010; Pham-Huy et al., 2008). In addition, oxidation reactions can also take place in meat products, especially for oxidation of lipids which play a main role in flavor formation (Ladikos and Lougovois, 1990). When oxidation of lipids is excessive, the undesirable physico-chemical characteristics of products including discoloration, drip loss, off-flavor, loss of nutrition and generation of toxic compounds are occurred (Falowo et al., 2014). There are several researches report that fermented meats are the potential sources of antioxidant peptides. The antioxidant activities of peptides presented in fermented meats are mainly due to proteolysis of muscle protein during fermentation and ripening process. Broncano et al. (2012) have studied the ability to quench free radical of low molecular weight compounds extracted from *Chorizo* sausages. The compounds including natural dipeptide carnosine together with Phe-Gly-Gly and L-carnitine which presented in hydrophilic fraction, were primarily responsible for the antioxidant activity of the *Chorizo* extract. Peptide fractions in dry-cured ham displayed the antioxidant properties which analyzed by the DPPH radical-scavenging assay and reducing power analysis. Peptide SAGNPN had the greatest radical-scavenging activity which was 50% antioxidant activity at a concentration of 1.5 mg/ml. In addition, the peptide GLAGA exhibited the higher reducing power with 0.5 units of absorbance at 700 nm at a concentration of 1 mg/ml. Spanish dry-cured ham was a source of antioxidant peptides naturally generated during the dry-curing process (Escudero et al., 2013). Fermented meat sauce presented high antioxidant activities against the DPPH radical and the OH-radical which were the values about 0.55 μmol Trolox eq./ml and 61.2%, respectively. Antioxidant activities of fermented meat sauce were related to proteolysis during fermentation and ripening. Fraction containing peptide Gln-Tyr-Pro and 2 kinds of unidentified substances were likely to contribute to high antioxidant activity in FMS (Ohata et al., 2016). Since some strains of LAB possed the important function such as antioxidant activities, several studies have applied LAB as probiotic and/or starter culture into fermented meats to enhance positive-health effect and quality of product. *Lb. plantarum* NBIMCC 2415 revealed comparatively high antioxidant activity against peroxide radicals and also contained the probiotic characteristics. The application of this strain as starter culture with antioxidant activity in fermented chicken sausages preserved color of the meat products (Nedelcheva et al., 2010). *Ped. pentosaceus*-inoculated dry sausages expressed remarkable reduction in the level of thiobarbituric acid-reactive substance, carbonyl formation and also decreased sulfhydryl loss in sausages. The reduction of volatile compounds that were responsible for lipid-oxidation including aldehydes, ketones and hydrocarbons, was presented in *Ped. pentosaceus*-inoculated dry sausages. Hence, *Ped. pentosaceus* had the potential to be employed as an antioxidant starter culture in fermented meat products (Chen et al., 2015). Low molecular weight nitrogen compounds and the antioxidant activity were markly affected by the microbial population in fermented sausages especially when starter cultures were employed. Starter cultures with high proteolytic activity helped to increase the antioxidant activity, mainly due to an increase in free amino acids and amines such as Lys, Met and tyramine, and also small peptides. The use of starters P200S34 (*Ped. acidilactici* M200 and *S. vitulus* RS34) can delay oxidative changes during storage of sausages (Fernández et al., 2016). *Lb. plantarum*-fermented chicken sausages (without spices) supplemented with dextrose and starch displayed higher antioxidant scavenging activity against hydroxyl, superoxide anion (SASA) and DPPH free radicals when compared to the treatment without addition of sugar. Furthermore, fermented chicken sausages supplemented with dextrose and starch had lower lipid oxidation, received better sensory scores and contained lower microbial count compared to the treatment without addition of sugar (Yadav et al., 2017). Zhang et al. (2017) evaluated the antioxidant activity of two strains of LAB including *Lb. curvatus* SR6 and *Lb. paracasei* SR10-1. They reported that *Lb. curvatus* SR6 exhibited better DPPH• scavenging activity (59.67% ± 6.68%) and reducing power (47.31% ± 4.62%) whereas, *Lb. paracasei* SR10-1 had better OH• scavenging activity (285.67% ± 2.00%) and anti-lipid peroxidation capacity (63.89% ± 0.93%). Both strains had the potential to use as antioxidant starters in sausage production due to their excellent antioxidant activities.

4.3. Angiotensin Converting Enzyme (ACE) Inhibitory Activity

Angiotensin I converting enzyme is a dipeptidyl carboxypeptidase [EC 3.4.15.1] that not only catalyzes conversion of angiotensin I to angiotensin II causing hypertension, but also inactivates bradykinin, a vasodilatory peptide (Erdös et al., 2010). The compounds that contain the ability to inactivate the action of this enzyme would decrease the blood pressure. Angiotensin I converting enzyme (ACE) inhibitory peptides have been derived from the hydrolysis of sarcoplasmic and myofibrillar porcine proteins during fermentation of meats by the role of lactic acid bacteria proteases (Castellano et al., 2013). Hence, fermented meats are the important sources of bioactive substances. The protein extracts derived from *Petrovac* sausage (traditional dry fermented sausage of Serbia) exhibited ACE inhibitory activity, which also increased; from 27.11 ± 2.163% (day 0) to 73.74 ± 3.299% (day 90). Protein hydrolysates produced during sausage ripening have a potent of ACE-inhibitory activity that may have positive effect on health (Vaštag et al., 2010). Protein hydrolysate obtained from the hydrolysis of sarcoplasmic protein by the action of *Lb. sakei* CRL1862 and *Lb. curvatus* CRL705 provided the peptide sequence FISNHAY which exhibited ACE inhibitory activity. The use of lactobacilli is beneficial for the development of functional meat products (Castellano et al., 2013). The addition of the protease Eg222 and starter culture P200S34 consisting *Ped. acidilactici* MS200 and *S. vitulus* RS34 in the processing of dry-fermented sausage *Salchichón* resulted in the highest ACE-inhibitory activity at 63 days of ripening which related to high nitrogen compound concentrations at the same day of ripening (Fernández et al., 2016a). According to the ability to degrade meat protein and tolerate salt and nitrite, *Lb. sakei* strain no. 23 and *Lb. curvatus* strain no. 28 were used as starter cultures in fermented sausages. The sausages fermented with these strains presented a great ACE inhibitory activity which contributed to fermentation products mainly bioactive peptides (Takeda et al., 2017). Low molecular weight peptides generated during meat fermentation not only contributed to flavor development but also exerted certain bioactivities. The factors that affected bacterial population, peptide concentration, and peptide size of dry-fermented sausages were ripening time and inoculated bacteria. When the ripening period extended, the antihypertensive capacity was increased. Protein fractions with peptides below 3 kDa showed the highest bioactive activity (Mejri et al., 2017). Gallego et al. (2018) have evaluated the changes in peptide profile and amino acid contents of three European dry-fermented sausages produced in Spain, Italy and Belgium. The great degradation of proteins, mainly myofibrillar and the generation of high amounts of different size peptides and free amino acids were occurred in these products. The water-soluble peptide extracts from Spanish and Belgian dry-fermented sausages showed values of ACE inhibition around 85%. Hence, dry-fermented sausages are found to be the natural sources of bioactive peptides. Fermented camel sausages had the higher proteolytic activities than those fermented beef sausages which resulted in greater ACE-inhibition capacity. The camel sausages fermented by the probiotic *Lb. plantarum* only had comparable proteolytic activities to those fermented by a commercial starter culture (SC only). Moreover, camel sausages fermented by *Lb. plantarum* only or SC + *Lb. plantarum* had greater antioxidant capacities, ACE-inhibition and cytotoxicity activity than those in beef sausage (Ayyash et al., 2019).

4.4. Antidiabetic Activity and Cytotoxicity

The antidiabetic activity and cytotoxicity of fermented meats have been infrequently investigated. Ayyash et al. (2020) have reported the comparison of health benefit capacity of fermented camel and beef sausages. Fermented camel and beef sausages inoculated with starter cultures and *Lc. lactis* KX881782 showed the greater antidiabetic activity than those inoculated with only starter cultures. Moreover, the fermented camel sausages exhibited lower antidiabetic activity than the beef sausages which was related to the nature of released peptide of camel sausages. The cytotoxicity or anti-cancer activity derived from peptides which contain the competing ability with cancer growth factors on the receptors of cancer cell-membrane and the ability to induce apoptosis or necrosis in cancer cell; or to their capability to inhibit gene transcription/cell proliferation. The cytotoxic activity against colon-cancer cell (Caco-2) of fermented camel sausages were greater than beef sausages. The cytotoxic activity against breast-cancer cell line (MCF-7) in camel sausages fermented by *Lc. lactis* was higher compared to other camel or beef sausages at 21-day of ripening. In addition, cytotoxic activity increased when the storage time was extended. Thus, *Lc. lactis* KX881782 is a promising probiotic culture with antidiabetic and cytotoxic potential to the fermented sausages.

5. Health Risks of Fermented Meats

5.1. Foodborne Pathogens

The consumption of foodborne pathogens contaminated in foodstuff is one of the main causes of several diseases (Ananchaipattana et al., 2012). The principal pathogens of concern are *Clostridium (Cl.) botulinum*, *Staphylococcus (S.) aureus*, *Listeria (L.) monocytogenes*, *Salmonella* (*Sal.*) spp., and *Escherichia (E.) coli*. These pathogens may influence the safe consumption of fermented meat (Carvalho et al., 2019).

Firstly, *Cl. botulinum* is defined as causative agent of botulism because it can produce botulinum neurotoxin that causes the severe neuroparalytic condition. Botulinum neurotoxin then gains access to the nervous system, where it obstructs the neurotransmitter acetylcholine releasing, preventing muscle contraction and thereby causing flaccid paralysis (Austin, 2014). The prevalence of *Cl. botulinum* in fermented meats has been reported (Schneider et al., 2011).

Secondly, *S. aureus* is one of the most prevalent pathogenic bacteria causing food-borne disease worldwide due to it can produce perilous toxin (Bergdoll and Lee Wong 2006). *S. aureus* is concerned for food products such as *Nham*, because the manufacturing process involves direct human handling, meat handling surfaces and the skin or hides of animals during processing (Petchsing and Woodburn, 1990). The study of Chokesajjawatee (2009) and Petchsing and Woodburn (1990) found and possibly concluded that *Nham* is frequently contaminated with *S. aureus*.

Thirdly, *L. monocytogenes* has been isolated from several ready-to-eat (RTE) food products including meat products (Meloni et al., 2014). Hence, the contamination of fermented sausages by *L. monocytogenes* would render their consumption and impact public health risk (Levine et al., 2001). Importantly, *L. monocytogenes* is able to survive in harsh condition of fermented meat products (Farber and Peterkin, 1991, Colville and Berryhill, 2007). *L. monocytogenes* is the major cause of listeriosis and the clinical manifestations of listeriosis are diverse such as flu-like symptoms, nausea and diarrhea (Stanner and Gibson-Moore, 2013).

Fourthly, the two *Salmonella enterica* serovars Typhimurium and Epidermidis are epidemic among human cases. *Salmonella* strains have been implicated in several outbreaks connected to consumption of dry-fermented sausages (DFSs). Reported outbreaks seem to be dominated by fermented sausages produced from pork meat contaminated with *Sal.* Typhimurium and other serovars (Holck et al., 2017; Kuhn et al., 2011; Bremer et al., 2004; Gossner et al., 2012). As for reduction of other pathogens, the using of potentially effective starter cultures has a greatly positive effect on *Salmonella* reductions (Coroller et al., 2015; Cenci-Goga et al., 2012).

Fifthly, *E. coli* O157:H7 (enterohaemorrhagic *E. coli*, EHEC) is a major and hazardous foodborne pathogen causing severe human disease (Lim et al., 2010). The prevalence of *E. coli* O157:H7 in fermented dry-cured salami has been reported. Noticeably, it is able to survive the process of fermented meat production and also it can adapt to acidic condition during fermentation. (Chacon et al., 2006). It produces Shiga-like toxin(s) that may cause abdominal cramps and bloody diarrhea that can progress through invasive infection to hemorrhagic colitis (HC), hemolytic uremic syndrome (HUS), and mortality (Berry and Well, 2010; Al-Nabuksi et al., 2016; Kivanc and Yapici, 2019).

Finally, the effective method to control and prevent the foodborne pathogens is the combination of utilization of potential starter, safety monitoring, good hygiene, and standard manufacturing.

5.2. Biogenic Amines and Nitrosamines

Biogenic amines (BAs) are normally found in fermented food products and synthesized from endogenous decarboxylases and the exogenous enzymes (microbial decarboxylation) (Tosukhowong et al., 2011; Kumar et al., 2017). BAs is double-edged sword. Because the BAs play key roles in human, animal, and plants physiology and the consumption of fermented food containing high level of BAs can be also causes of the several serious human health risks (Sivamaruthi et al., 2019; Bogdanović et al., 2020). Adverse effects of BAs (i.e., histamine and tyramine) include gastric, headaches, intestinal problems, migraine, and pseudoallergic responses and BAs are also considered as the nitrosamine carcinogenic precursors (ten Brink et al., 1990; Karovičová and Kohajdová, 2005).

In dry-fermented sausages, the BAs are tyramines, cadaverine, putrescine and histamine (Eerola et al., 1996). The dominate BAs in pork meat marinade found were putrescine, cadaverine, and phenylethylamine (Mozuriene et al., 2016). The study of Bover-Cid and Holzapfel (1999) found that the high levels of tyramine and putrescine and production of cadaverine, agmatine, tryptamine, and phenylethylamine were found in dry fermented sausages. In Thai traditional fermented meat (*Nham*), prior to *Nham* processing, pork meat was kept at 30 °C for 6 h, and at 4 and -20 °C for 2 days. The formation of biogenic amines (cadaverine, putrescine, histamine, and tyramine) in *Nham*, which is processed from stored meat, was significantly and comparatively higher than *Nham*, which processed from fresh meat. Interestingly, the attractive strategy is the using of potentially commercial starter (*Lb. plantarum* BCC 9546) because this strain can significantly decrease the levels of these BAs during the process of Nham fermentation (Tosukhowong et al., 2011). Importantly, the BAs formation can be regulated by the using of appropriate starter culture, efficient production, and standard hygiene (Kumar et al., 2017).

For nitrosamines, nitrate/nitrite is frequently supplemented into dry fermented sausage manufacture because it plays key role in color and flavor development and their anti-oxidant activity, and it can also inhibit the growth of spoilage microorganisms (*Cl. botulinum*) (Shimokomaki et al., 2003). Despite the beneficial role of nitrate and nitrite in fermented products, there is health concern regarding the role of these additives in the formation of nitrosamines, which carcinogenic, mutagenic and teratogenic effects are well documented (Honikel, 2008; Hospital et al., 2012). Nitrosamines are formed during a reaction of nitrite or nitrate with certain amines and they have been found to be carcinogenic substance. (Bauer, 2014). In the mouth or stomach, human can also convert nitrate to carcinogenic nitrosamines if the food comprises of nitrosamine precursors (Robles, 2014). In meat products, nitrosopiperidine and nitrosopyrrolidine are important (Bauer, 2014). Thus, some fermented meat products use natural curing agents and avoid using the nitrite compound in high amount. The study of Cenci-Goga et al. (2012) reported that the using of LAB starter in the production of nitrite-free low-acid fermented venison (*Dama dama*) sausage (salame di daino) can inhibit and decrease the counts of *Enterobacteriaceae*, coliform organisms and *Pseudomonas* spp. Perea-Sanz et al. (2020) reported that in dry fermented sausage, the *Debaryomyces hansenii* inoculation can reduce nitrite/nitrate levels. Moreover, the findings of Jin et al. (2018) found that the celery powder was the naturally potent curing agents which can protect sausages from quality deterioration during storage. Thus, nitrosamine compounds in fermented meat products is the important things that must be limited and controlled.

5.3. Mycotoxin

Mycotoxins are a group of toxic compounds produced by molds. Several mycotoxins (aflatoxins and ochratoxin A) were reported to be present in fermented foods because mycotoxins can enter the food supply indirectly (using of contaminated ingredients in food products) and directly (contamination of grains and stored products) (Ames, 1989; Sivamaruthi et al., 2019; Álvarez et al., 2019).

Aflatoxins (AF) are mainly produced by the species of *Aspergillus* (i.e., *A. flavus* and *A. parasiticus*) (Tirmenstein and Mangipudy, 2014) and distributed in fermented meat products (Markov et al., 2013; Pleadin et al., 2015). This toxin affects adversely to human health due to it can be severely potential immunosuppressant, hepatotoxins, mutagens and carcinogens (Coppock and Christian, 2007). Importantly, the International Agency for Research on Cancer (IARC) reported that AF is categorized as group I carcinogens (Human and Cancer, 2002).

Ochratoxin A (OA) is an important foodborne toxin and mainly produced by species of *Aspergillus* (*A.*) and *Penicillium* (*P.*) (Duarte et al., 2010). OA has a number of adverse effects, such as nephrotoxicity, hepatotoxicity and carcinogenicity (Wilson et al., 2016). Importantly, OA has also been found in fermented meat products (Markov et al., 2013; Vulić et al., 2016).

Noticeably, the attractive reduction, control, and prevention approaches of mycotoxin exposure in food are the using of effective starter and/or protective culture, biological anti-fungal and ingredient active compound, and great standard procedures (from farm, transportation until factory, distribution and preservation). For example, the study of Álvarez et al. (2019) reported that in dry-fermented sausages, the using of *En. faecium* SE920 (protective culture) can inhibit the growth of *P. nordicum;* consequently, it can also significantly decrease the OA production from *P. nordicum*. Delgado et al. (2015) found that the using of anti-fungal protein which produced from *P. chrysogenum* (PgAFP) can inhibit unwanted molds, including the main mycotoxin-producing species of *Aspergillus* and *Penicillium* of concern for dry-ripened foods. The findings of Martín-Sánchez et al. (2011) reported that in Spanish fermented dry-cured sausages (*Salchichón*), the using of oregano essential oil (OEO) can inhibit surface mold. Moreover, Andrade et al. (2019) reported that the replacement of NaCl with KCl and sucrose minimized the presence of AF and OA accumulation in cured meat products.

5.4. Polycyclic Aromatic Hydrocarbons (PAHs)

Polycyclic aromatic hydrocarbons (PAHs) are a group of organic compounds, which are composed of two or more fused aromatic rings. Most of PAHs are mutagens and carcinogens (Fan and Lin, 2011). Thus, the long-term exposure of PAHs and the consumption of meat product contaminating with PAHs could lead to several disorders such as mutations, developmental malformations, tumors, and cancer. PAHs are mainly obtained from incomplete combustion of coal and oil such as smoking food processing (Fretheim, 1976: Wu, 2020). The benzo[a]pyrene (BaP) is one the members of PAHs and its concentration has received attention. Because its higher contribution to overall burden of cancer in humans, also being used as a marker for the occurrence and effect of carcinogenic PAHs in food (Rey-Salgueiro et al., 2009). Gomes et al. (2013) reported that fat contents and smoking procedure did not affect on PAHs contents while casing type was the most impact factor and significantly affected on PAHs contents in Portuguese traditional dry fermented sausages.

Interestingly, LAB can play beneficial role in the reduction of PAHs accumulation in fermented meat products. The study Bartkiene et al. (2017) reported that the treatment of sausages surface with LAB before and after smoking significantly reduced both benzo[a]pyrene and chrysene. Škaljac et al. (2014) reported that the indirect smoking procedure was safe, and this procedure can reduce the PAHs formation in dry Serbia fermented sausages (*Petrovská klobása*). Thus, the starter culture, the using of LAB treatment, cooking procedures, the selection ingredient are the important key of prevention, reduction avoidance PAHs formation.

CONCLUSION

Fermented meats are the sources of lactic acid bacteria and coagulase-negative staphylococci which are selected to the beneficial tools for enhancing quality, sensory attributes and safety of the products. Fermented meats are consumed not only for nutritional values, but also for physiological functions on health. Bioactive peptides and free amino acids containing in fermented meats exert antioxidant, ACE Inhibitory, antidiabetic activity and cytotoxicity which provide the positive effects on health. Moreover, the application of probiotic cultures also exhibits health functionality due to their probiotic properties. However, the nonstandard and unsuitable processing and poor-quality ingredients are the major cause of several health risks. The fermented meat contaminating with undesirable compounds and some pathogens can make adverse effects to consumer. Thus, the good sanitation processing and the appropriate consumption of fermented products are the necessarily consideration.

REFERENCES

Adams, M. R. 1986. Fermented flesh foods. *Progress in industrial microbiology*, 23:159-198.

Adewumi, G. A. 2019. "Health-Promoting Fermented Foods." In: Melton, L., Shahidi, F., Varelis, P. (Eds.), *Encyclopedia of Food Chemistry*, vol. 3, pp. 399–418. Elsevier.

Al-Nabulsi, A. A., Olaimat, A. N., Osaili, T. M., Ayyash, M. M., Abushelaibi, A., Jaradat, Z. W. Shaker, R., Al-Taani, M. and Holley, R. A. 2016. "Behavior of *Escherichia coli* O157: H7 and Listeria monocytogenes during fermentation and storage of camel yogurt." *Journal of Dairy Science* 99 (3):1802-1811.

Álvarez, M., Rodríguez, A., Peromingo, B., Núñez, F. and Rodríguez, M. 2019. "*Enterococcus faecium*: A promising protective culture to control growth of ochratoxigenic moulds and mycotoxin production in dry-fermented sausages." *Mycotoxin Research*:1-9.

Ames, I. A. 1989. "*Mycotoxins, economic and health risks*." CAST (Council of Agricultural Science and Technology), Task force Rep. 116.

Ananchaipattana, C., Hosotani, Y., Kawasaki, S., Pongsawat, S., Md. Latiful, B., Isobe, S. and Inatsu, Y. 2012. "Prevalence of foodborne pathogens in retailed foods in Thailand." *Foodborne pathogens and disease* 9 (9):835-840.

Andrade, M. J, Peromingo, B., Rodríguez, M. and Rodríguez, A. 2019. "Effect of cured meat product ingredients on the *Penicillium verrucosum* growth and ochratoxin A production." *Food Control* 96:310-317.

Austin, J. W. 2014. "*Clostridium botulinum and Botulism*." In, 330-334.

Ayyash, M., Liu, S. -Q., Al Mheiri, A., Aldhaheri, M., Raeisi, B., Al-Nabulsi, A., Osailic, T. and Olaimat, A. 2019. *In vitro* investigation of health-promoting benefits of fermented camel sausage by novel probiotic *Lactobacillus plantarum*: A comparative study with beef sausages. *LWT - Food Science and Technology*, 99: 346–354.

Ayyash, M., Olaimat, A., Al-Nabulsi, A. and Liu, S. -Q. 2020. Bioactive properties of novel probiotic *Lactococcus lactis* fermented camel sausages: cytotoxicity, angiotensin converting enzyme inhibition, antioxidant capacity, and antidiabetic activity. *Food Science of Animal Resources*, 40(2): 155–171.

Ba, H. V., Seo, H. -W., Seong, P. -N., Kang, S. -M., Kim, Y. -S., Cho, S. -H., Park, B. -Y., Ham, J. -S. and Kim, J. -H. 2018. *Lactobacillus plantarum* (KACC 92189) as a potential probiotic starter culture for quality improvement of fermented sausages. *Korean Journal for Food Science of Animal Resources*, 38(1): 189–202.

Babić, I., Markov, K., Kovačević, D., Trontel, A., Slavica, A., Đugum, J., Čvek, D., Svetec, I. K., Posavec, S. and Frece, J. 2011. Identification and characterization of potential autochthonous starter cultures from a Croatian "brand" product "Slavonski kulen." *Meat Science*, 88: 517–524.

Bartkiene, E., Bartkevics, V., Mozuriene, E., Krungleviciute, V., Novoslavskij, A., Santini, A., Rozentale, I., Juodeikiene, G. and Cizeikiene, D. 2017. "The impact of lactic acid bacteria with antimicrobial properties on biodegradation of polycyclic aromatic hydrocarbons and biogenic amines in cold smoked pork sausages." *Food Control* 71:285-292.

Bauer, F. 2014. "*Residues in meat and meat products| residues associated with meat production.*"

Bergdoll, M. S. and Wong, A. C. L. 2006. "Staphylococcal intoxications. 523-562." *Foodborne Infections and Intoxications,* Academic Press, Elsevier Inc., San Diego, California.

Berry, E. D. and James, E. Wells. 2010. "Chapter 4 - *Escherichia coli* O157:H7: Recent Advances in Research on Occurrence, Transmission, and Control in Cattle and the Production Environment." In *Advances in*

Food and Nutrition Research, edited by Steve L. Taylor, 67-117. Academic Press.

Bogdanović, T., Petričević, S., Brkljača, M., Listeš, I. and Pleadin, J. 2020. "Biogenic amines in selected foods of animal origin obtained from the Croatian retail market." *Food Additives & Contaminants: Part A* 37 (5):815-830.

Bover-Cid, S. and Holzapfel, W. H. 1999. "Improved screening procedure for biogenic amine production by lactic acid bacteria." *International journal of food microbiology* 53 (1):33-41.

Bremer, V., Leitmeyer, K., Jensen, E., Metzel, U., Meczulat, H., Weise, E., Werber, D., Tschaepe, H., Kreienbrock, L. and Glaser, S. 2004. "Outbreak of Salmonella Goldcoast infections linked to consumption of fermented sausage, Germany 2001." *Epidemiology & Infection* 132 (5):881-887.

Broncano, J. M., Otte, J., Petrón, M. J., Parra, V. and Timón, M. L. 2012. Isolation and identification of low molecular weight antioxidant compounds from fermented "chorizo" sausages. *Meat Science*, 90: 494–501.

Bungenstock, L., Abdulmawjood, A. and Reich, F. 2020. Evaluation of antibacterial properties of lactic acid bacteria from traditionally and industrially produced fermented sausages from Germany. *PLoS ONE*, 15(3): e0230345.

Campbell-Platt, G. and Cook, P. E. *Fermented Meat, Blackie Academic and Professional*, London, 1995; pp. 242.

Campos, S. D., Alves, R. C., Mendes, E., Costa, A. S. G., Casal, S. and Oliveira, M. B. P. P. 2013. Nutritional value and influence of the thermal processing on a traditional Portuguese fermented sausage (*alheira*). *Meat Science*, 93: 914–918.

Cardinali, F., Milanovića, V., Osimani, A., Aquilanti, L., Taccari, M., Garofalo, C., Polverigiani, S., Clementi, F., Franciosi, E., Tuohy, K., Mercuri, M. L., Altissimi, M. S. and Haouet, M. N. Microbial dynamics of model Fabriano-like fermented sausages as affected by starter cultures, nitrates and nitrites. *International Journal of Food Microbiology*, 278: 61–72.

Carvalho, M. I., Albano, H. C. and Teixeira, P. 2019. "Influence of oregano essential oil on the inhibition of selected pathogens in "Alheira" during storage." *Acta Scientiarum Polonorum Technologia Alimentaria* 18 (1):13-23.

Castellano, P., Aristoy, M. C., Sentandreu, M. Á., Vignolo, G. and Toldrá, F. 2013. Peptides with angiotensin I converting enzyme (ACE) inhibitory activity generated from porcine skeletal muscle proteins by the action of meat-borne *Lactobacillus*. *Journal of Proteomics*, 89: 183–190.

Cenci-Goga, B. T., Rossitto, P. V., Sechi, P., Parmegiani, S., Cambiotti, V. and Cullor, J. S. 2012. "Effect of selected dairy starter cultures on microbiological, chemical and sensory characteristics of swine and venison (Dama dama) nitrite-free dry-cured sausages." *Meat science* 90 (3):599-606.

Chacon, P. A., Muthukumarasamy, P. and Holley, R. A. 2006. "Elimination of *Escherichia coli* O157: H7 from fermented dry sausages at an organoleptically acceptable level of microencapsulated allyl isothiocyanate." *Applied and environmental microbiology* 72 (5):3096-3102.

Charmpi, C., Van der Veken, D., Van Reckem, E., De Vuyst, L. and Leroy, F. 2020. Raw meat quality and salt levels affect the bacterial species diversity and community dynamics during the fermentation of pork mince. *Food Microbiology*, 89: 103434.

Chen, Q., Kong, B., Sun, Q., Dong, F. and Liu, Q. 2015. Antioxidant potential of a unique LAB culture isolated from Harbin dry sausage: *In vitro* and in a sausage model. *Meat Science*, 110: 180–188.

Chikthimmah, N. and Knabel, S. J. 2001. "Survival of *Escherichia coli* O157: H7, *Salmonella* Typhimurium, and *Listeria monocytogenes* in and on vacuum packaged Lebanon bologna stored at 3.6 and 13.0 C." *Journal of food protection* 64 (7):958-963.

Colville, J. L. and Berryhill, D. L. 2007. "LISTERIOSIS." In *Handbook of Zoonoses*, edited by Joann L. Colville and David L. Berryhill, 108-113. Saint Louis: Mosby.

Coppock, R. W. and Ralph, G. Christian. 2007. "CHAPTER 75 - Aflatoxins." In *Veterinary Toxicology*, edited by Ramesh C. Gupta, 939-950. Oxford: Academic Press.

Coroller, L., Jeuge, S., Couvert, O., Christieans, S. and Ellouze, M. 2015. "Extending the gamma concept to non-thermal inactivation: a dynamic model to predict the fate of *Salmonella* during the dried sausages process." *Food microbiology* 45:266-275.

Curtui, V. G., Gareis, M., Usleber, E. and Märtlbauer, E. 2001. "Survey of Romanian slaughtered pigs for the occurrence of mycotoxins ochratoxins A and B, and zearalenone." *Food Additives & Contaminants* 18 (8):730-738.

Delgado, J., Acosta, R., Rodríguez-Martín, A., Bermúdez, E., Núñez, F. and Asensio, M. A. 2015. "Growth inhibition and stability of PgAFP from *Penicillium chrysogenum* against fungi common on dry-ripened meat products." *International Journal of Food Microbiology* 205:23-29.

Duarte, S. C., Pena, A. and Lino, C. M. 2010. "Ochratoxin A in Portugal: a review to assess human exposure." *Toxins* 2 (6):1225-1249.

Eerola, S., Maijala, R., SAGUÉS, A. X. R., Salminen, M. and Hirvi, T. 1996. "Biogenic amines in dry sausages as affected by starter culture and contaminant amine-positive Lactobacillus." *Journal of Food Science* 61 (6):1243-1246.

Erdös, E. G., Tan, F. and Skidgel, R. A. 2010. Angiotensin I-converting enzyme inhibitors are allosteric enhancers of kinin B1 and B2 receptor function. *Hypertension*, 55: 214–20.

Escudero, E., Mora, L., Fraser, P. D., Aristoy, M. -C. and Toldrá, F. 2013. Identification of novel antioxidant peptides generated in Spanish dry-cured ham. *Food Chemistry*, 138: 1282–1288.

Falowo, A. B., Fayemi, P. O. and Muchenje, V. 2014. Natural antioxidants against lipid-protein oxidative deterioration in meat and meat products: A review. *Food Research International*, 64: 171–181.

Fan, Z. and Lin, L. 2011. "Exposure Science: Contaminant Mixtures." In *Encyclopedia of Environmental Health*, edited by J. O. Nriagu, 645-656. Burlington: Elsevier.

FAO/WHO. (2002). *Guidelines for the evaluation of probiotics in food, Food and Agriculture Organization of the United Nations and World Health Organization Working Group Report*, London, Ontario: 1-11.

Fernández, M., Benito, M. J., Martín, A., Casquete, R., Córdoba, J. J. and Córdoba, M. G. 2016a. Influence of starter culture and a protease on the generation of ACE-inhibitory and antioxidant bioactive nitrogen compounds in Iberian dry-fermented sausage "*salchichón*." *Heliyon*, 2: e00093.

Fernández, M., Martín, A., Benito, M. J., Casquete, R., Recio, I. and De Guía Córdoba, M. 2016b. Influence of starter cultures on the generation of antioxidant nitrogen compounds in Iberian dry-fermented sausages. *International Journal of Food Science and Technology*, 51: 435–443.

Fontana, C., Bassi, D., López, C., Pisacane, V., Otero, M. C., Puglisi, E., Rebecchi, A., Cocconcelli, P. S. and Vignolo, G. 2016. Microbial ecology involved in the ripening of naturally fermented llama meat sausages. A focus on lactobacilli diversity. *International Journal of Food Microbiology*, 236: 17–25.

Franco, D., Martins, A. J., López-Pedrouso, M., Cerqueira, M. A., Purriños, L., Pastrana, L. M., Vicente, A. A., Zapata C. and Lorenzo, J. M. 2020. Evaluation of linseed oil oleogels to partially replace pork backfat in fermented sausages. *Journal of the Science of Food and Agriculture*, 100: 218–224.

Fretheim, K. 1976. "Carcinogenic polycyclic aromatic hydrocarbons in Norwegian smoked meat sausages." *Journal of agricultural and food chemistry* 24 (5):976-979.

Gallego, M., Mora, L., Escudero, E. and Toldrá, F. 2018. Bioactive peptides and free amino acids profiles in different types of European dry-fermented sausages. *International Journal of Food Microbiology*, 276: 71–78.

Gomes, A., Santos, C., Almeida, J., Elias, M. and Roseiro L. C. 2013. "Effect of fat content, casing type and smoking procedures on PAHs contents of Portuguese traditional dry fermented sausages." *Food and chemical toxicology* 58:369-374.

Gossner, C. M., Van Cauteren, D., Le Hello, S., Weill, F. X., Terrien, E., Tessier, S., Janin, C., Brisabois, A., Dusch, V. and Vaillant, V. 2012. "Nationwide outbreak of *Salmonella* enterica serotype 4,[5], 12: i:- infection associated with consumption of dried pork sausage, France, November to December 2011." *Eurosurveillance* 17 (5):20071.

Holck, A., Axelsson, L., McLeod, A., Rode, T. M. and Heir, E. 2017. "Health and safety considerations of fermented sausages." *Journal of Food Quality* 2017.

Holko, I., Hrabě, J., Šalaková, A. and Rada, V. 2013. The substitution of a traditional starter culture in mutton fermented sausages by *Lactobacillus acidophilus* and *Bifidobacterium animalis*. *Meat Science*, 94: 275–279.

Honikel, K. O. 2008. "The use and control of nitrate and nitrite for the processing of meat products." *Meat science* 78 (1-2):68-76.

Hospital, X. F., Hierro, E. and Fernández, M. 2012. "Survival of *Listeria innocua* in dry fermented sausages and changes in the typical microbiota and volatile profile as affected by the concentration of nitrate and nitrite." *International Journal of Food Microbiology* 153 (3):395-401.

Hospital, X. F., Hierro, E. and Fernandez, M. 2014. "Effect of reducing nitrate and nitrite added to dry fermented sausages on the survival of *Salmonella* Typhimurium." *Food research international* 62:410-415.

Humans, IARC Working Group on the Evaluation of Carcinogenic Risks to Humans and International Agency for Research on Cancer. 2002. *Some traditional herbal medicines, some mycotoxins, naphthalene and styrene*: World Health Organization.

Iacumina, L., Osualdini, M., Bovolenta, S., Boscolo, D., Chiesa, L., Panseri, S. and Comi, G. 2020. Microbial, chemico-physical and volatile aromatic compounds characterization of *Pitina* PGI, a peculiar sausage-like product of North East Italy. *Meat Science*, 163: 108081.

Jiménez-Colmenero, F., Triki, M., Herrero, A. M., Rodríguez-Salas, L. and Ruiz-Capillas, C. 2013. Healthy oil combination stabilized in a konjac matrix as pork fat replacement in low-fat, PUFA-enriched, dry fermented sausages. *LWT - Food Science and Technology*, 51: 158–163.

Jin, S. K., Choi, J. S., Yang, H. S., Park, T. S. and Yim, D. G. 2018. "Natural curing agents as nitrite alternatives and their effects on the physicochemical, microbiological properties and sensory evaluation of sausages during storage." *Meat science* 146:34-40.

Josquin, N. M., Linssen, J. P. H. and Houben, J. H. 2012. Quality characteristics of Dutch-style fermented sausages manufactured with partial replacement of pork back-fat with pure, pre-emulsified or encapsulated fish oil. *Meat Science*, 90: 81–86.

Kargozari, M., Moini, S., Basti, A. A., Emam-Djomeh, Z., Ghasemlou, M., Martin, I. R., Gandomi, H., Carbonell-Barrachina, A. A. and Szumny, A. 2014. Development of Turkish dry-fermented sausage (sucuk) reformulated with camel meat and hump fat and evaluation of physicochemical, textural, fatty acid and volatile compound profiles during ripening. *LWT - Food Science and Technology*, 59: 849–858.

Karovičová, J. and Kohajdová, Z. 2005. "Biogenic amines in food." *Chemical papers* 59 (1):70-79.

Kivanc, M. and Yapici, E. 2019. "Survival of *Escherichia coli* O157: H7 and *Staphylococcus aureus* during the fermentation and storage of kefir." *Food Science and Technology* 39:225-230.

Klingberg, T. D., Axelsson, L., Naterstad, K., Elsser, D. and Budde, B. B. 2005. Identification of potential probiotic starter cultures for Scandinavian-type fermented sausages. *International Journal of Food Microbiology* 105, 419– 431.

Kuhn, K. G., Torpdahl, M., Frank, C., Sigsgaard, K. and Ethelberg, S. 2011. "An outbreak of *Salmonella* Typhimurium traced back to salami, Denmark, April to June 2010." *Eurosurveillance* 16 (19):19863.

Kumar, P., Chatli, M. K., Verma, A. K., Mehta, N., Malav, O.P., Kumar, D. and Sharma, N. 2017. "Quality, functionality, and shelf life of fermented meat and meat products: A review." *Critical reviews in food science and nutrition* 57 (13):2844-2856.

Ladikos, D. and Lougovois, V. 1990. Lipid oxidation in muscle foods: A review. *Food Chemistry*, 35: 295–314.

Leroy, F., Verluyten, J. and De Vuyst, L. (2006). Functional meat starter cultures for improved sausage fermentation. *Int. J. Food Microbiol.* 106:270–285.

Levine, P., Rose, B., Green, S., Ransom, G. and Hill, W. 2001. "Pathogen testing of ready-to-eat meat and poultry products collected at federally inspected establishments in the United States, 1990 to 1999." *Journal of Food Protection* 64 (8):1188-1193.

Lim, J. Y., Yoon, J. W. and Hovde, C. J. 2010. "A brief overview of *Escherichia coli* O157: H7 and its plasmid O157." *Journal of microbiology and biotechnology* 20 (1):5.

Liu, S. N., Han, Y., and Zhou, Z. J. 2011. Lactic acid bacteria in traditional fermented Chinese foods. *Food Research International* 44, 643–651.

Lucke, F. K., 1994. Fermented meat-products. *Food Research International* 27, 299–307.

Lucke, F. K. 1998. Fermented sausages. In: *Microbiology of Fermented Foods,* 2nd ed., Vol. 2, pp. 441–483. Wood, B. J. B., Ed., Blackie Acad. Prof., London.

Lv, J., Li, C., Li, S., Liang, H., Ji, C., Zhu, B. and Lin, X. 2019. Effects of temperature on microbial succession and quality of sour meat during fermentation. *LWT - Food Science and Technology*, 114: 108391.

Maksimovic, A. Z., Zunabovic-Pichler, M., Kos, I., Mayrhofer, S., Hulak, N., Domig, K. J. and Fuka, M. M. 2018. Microbiological hazards and potential of spontaneously fermented game meat sausages: A focus on lactic acid bacteria diversity. *LWT - Food Science and Technology*, 89: 418–426.

Markov, K., Pleadin, J., Bevardi, M., Vahčić, N., Sokolić-Mihalak, D. and Frece, J. 2013. "Natural occurrence of aflatoxin B1, ochratoxin A and citrinin in Croatian fermented meat products." *Food Control* 34 (2):312-317.

Martín-Sánchez, A. M., Chaves-López, C., Sendra, E., Sayas, E., Fenández-López, J. and Pérez-Álvarez, J. Á. 2011. "Lipolysis, proteolysis and sensory characteristics of a Spanish fermented dry-cured meat product (salchichón) with oregano essential oil used as surface mold inhibitor." *Meat science* 89 (1):35-44.

Marty, E., Buchs, J., Eugster-Meier, E., Lacroix, C. and Meile, L. 2012. Identification of staphylococci and dominant lactic acid bacteria in spontaneously fermented Swiss meat products using PCR-RFLP. *Food Microbiology*, 29: 157–166.

Mejri, L., Vásquez-Villanueva, R., Hassouna, M., Marina, M. L. and García, M. C. 2017. Identification of peptides with antioxidant and antihypertensive capacities by RP-HPLC-Q-TOF-MS in dry fermented camel sausages inoculated with different starter cultures and ripening times. *Food Research International*, 100: 708–716.

Meloni, D., Consolati, S. G., Mazza, R., Mureddu, A., Fois, F., Piras, F. and Mazzette, R. 2014. "Presence and molecular characterization of the major serovars of *Listeria monocytogenes* in ten Sardinian fermented sausage processing plants." *Meat science* 97 (4):443-450.

Menegas, L. Z., Pimentel, T. C., Garcia, S. and Prudencio, S. H. Effect of adding inulin as a partial substitute for corn oil on the physicochemical and microbiological characteristics during processing of dry-fermented chicken sausage. *Journal of Food Processing and Preservation*, 41: e13166.

Mozuriene, E., Bartkiene, E., Krungleviciute, V., Zadeike, D., Juodeikiene, G., Damasius, J. and Baltusnikiene, A. 2015. "Effect of natural marinade based on lactic acid bacteria on pork meat quality parameters and biogenic amine contents." *LWT-Food science and technology* 69:319-326.

Muguerza, E., Gimeno, O., Ansorena, D., Bloukas, J. G. and Astiasarán, I. 2001. Effect of replacing pork backfat with pre-emulsifiedolive oil on lipid fraction and sensory quality of Chorizo de Pamplona - a traditional Spanish fermented sausage. *Meat Science*, 59: 251–258.

Muthukumarasamy, P. and Holley, R. A. 2006. Microbiological and sensory quality of dry fermented sausages containing alginate-microencapsulated *Lactobacillus reuteri*. *International Journal of Food Microbiology*, 111: 164–169.

Nedelcheva, P., Denkova, Z., Denev, P., Slavchev, A. and Krastanov, A. 2010. Probiotic strain *Lactobacillus Plantarum* NBIMCC 2415 with antioxidant activity as a starter culture in the production of dried fermented meat. *Biotechnology & Biotechnological Equipment*, 24(1): 1624–1630.

Nguyen, D. T. L., Van Hoorde, K., Cnockaert, M., De Brandt, E., De Bruyne, K., Le, B. T. and Vandamme, P. 2013. A culture-dependent and -independent approach for the identification of lactic acid bacteria associated with the production of nem chua, a Vietnamese fermented meat product. *Food Research International*, 50: 232–240.

Nightingale, K. K., Thippareddi, H., Phebus, R. K., Marsden, J. L. and Nutsch, A. L. 2006. "Validation of a traditional Italian-style salami manufacturing process for control of *Salmonella* and *Listeria monocytogenes*." *Journal of food protection* 69 (4):794-800.

Nissen, H. and Holck, A. 1998. "Survival of *Escherichia coli* O157: H7, *Listeria monocytogenes* and *Salmonella* kentuckyin Norwegian fermented, dry sausage." *Food Microbiology* 15 (3):273-279.

Ohata, M., Uchida, S., Zhou, L. and Arihara, K. 2016. Antioxidant activity of fermented meat sauce and isolation of an associated antioxidant peptide. *Food Chemistry*, 194: 1034–1039.

Pavli, F. G., Argyri, A. A., Chorianopoulos, N. G., Nychas, G. -J. E. and Tassou, C. C. 2020. Effect of *Lactobacillus plantarum* L125 strain with probiotic potential on physicochemical, microbiological and sensorial characteristics of dry-fermented sausages. *LWT - Food Science and Technology*, 118: 108810.

Perea-Sanz, L., López-Díez, J. J., Belloch, C. and Flores, M. 2020. "Counteracting the effect of reducing nitrate/nitrite levels on dry fermented sausage aroma by Debaryomyces hansenii inoculation." *Meat Science*: 108103.

Petchsing, U. and Woodburn, M. J. 1990. "*Staphylococcus aureus* and *Escherichia coli* in nham (Thai-style fermented pork sausage)." *International journal of food microbiology* 10 (3-4):183-192.

Phalakornkule, C. and Tanasupawat, S. 2006. "Characterization of Lactic Acid Bacteria from Traditional Thai Fermented Sausages." *Journal of Culture Collections* (ISSN: 1310-8360) Vol 5 Num 1.

Pham-Huy, L. A., He, H. and Pham-Huy, C. (2008). Free radicals, antioxidants in disease and health. *International Journal of Biomedical Science*, 4: 89–96.

Pisacane, V., Callegari, M. L., Puglisi, E., Dallolio, G. and Rebecchi, A. Microbial analyses of traditional Italian salami reveal microorganisms transfer from the natural casing to the meat matrix. *International Journal of Food Microbiology*, 207: 57–65.

Pleadin, J., Kovačević, D. and Perković, I. 2015. "Impact of Casing Damaging on Aflatoxin B1 Concentration during the Ripening of Dry-Fermented Meat Sausages." *Journal of Immunoassay and Immunochemistry* 36 (6):655-666.

Pogorzelska-Nowicka, E., Atanasov, A. G., Horbańczuk, J. and Wierzbicka, A. 2018. Bioactive compounds in functional meat products. *Molecules*, 23: 307.

Prado, N., Sampayo, M., González, P., Lombó, F. and Díaz, J. Physicochemical, sensory and microbiological characterization of *Asturian Chorizo*, a traditional fermented sausage manufactured in Northern Spain. *Meat Science*, 156: 118–124.

Pringsulaka, O., Thongngam, N., Suwannasai, N., Atthakor, W., Pothivejkul, K. and Rangsiruji, A. 2012. Partial characterisation of bacteriocins produced by lactic acid bacteria isolated from Thai fermented meat and fish products. *Food Control*, 23: 547–551.

Quijada, N. M., De Filippis, F., Sanz, J. J., García-Fernández, M. d. C., Rodríguez-Lázaro, D., Ercolini, D. and Hernández, M. 2018. Different *Lactobacillus* populations dominate in "Chorizo de Léon" manufacturing performed in different production plants. *Food Microbiology*, 70: 94–102.

Rey-Salgueiro, L., Martínez-Carballo, E., García-Falcón, M. S., González-Barreiro, C. and Simal-Gándara, J. 2009. "Occurrence of polycyclic aromatic hydrocarbons and their hydroxylated metabolites in infant foods." *Food chemistry* 115 (3):814-819.

Robles, H. 2014. "Nitrosamines." In *Encyclopedia of Toxicology* (Third Edition*)*, edited by Philip Wexler, 584-585. Oxford: Academic Press.

Romero, M. C., Romero, A. M., Doval, M. M. and Judis, M. A. 2013. Nutritional value and fatty acid composition of some traditional Argentinean meat sausages. *Food Science and Technology*, 33(1): 161–166.

Rubio, R., Jofré, A., Aymerich, T., Guàrdia, M. D. and Garriga, M. 2014a. Nutritionally enhanced fermented sausages as a vehicle for potential probiotic lactobacilli delivery. *Meat Science*, 96: 937–942.

Rubio, R., Martín, B., Aymerich T. and Garriga, M. 2014b. The potential probiotic *Lactobacillus rhamnosus* CTC1679 survives the passage through the gastrointestinal tract and its use as starter culture results in safe nutritionally enhanced fermented sausages. *International Journal of Food Microbiology*, 186: 55–60.

Ruiz-Moyano, S., Martín, A., Benito, M. J., Hernández, A., Casquete R. and de Guía Córdoba, M. 2010. Safety and functional aspects of pre-selected pediococci for probiotic use in Iberian dry-fermented sausages. *International Journal of Food Science and Technology*, 45: 1138–1145.

Rzepkowska, A., Zielińska, D., Ołdak A. and Kołożyn-Krajewska, D. 2017. Safety assessment and antimicrobial properties of the lactic acid bacteria strains isolated from polish raw fermented meat products. *International Journal of Food Properties*, 20(11): 2736–2747.

Saarela, M., Mogensen, G., Fondén, R., Mättö, J. and Mattila-Sandholm, T. (2000). Probiotic bacteria: safety, functional and technological properties. *Journal of Biotechnology*, 84: 197–215.

Santiyanont, P., Chantarasakha, K., Tepkasikul, P., Srimarut, Y., Mhuantong, W., Tangphatsornruang, S., Zo, Y. -G. and Chokesajjawatee, N. 2019. Dynamics of biogenic amines and bacterial communities in a Thai fermented pork product Nham. *Food Research International*, 119: 110–118.

Santos, S. C., Fraqueza, M. J., Elias, M., Barreto, A. S. and Semedo-Lemsaddek, T. 2017. Traditional dry smoked fermented meat sausages: Characterization of autochthonous enterococci. *LWT - Food Science and Technology*, 79:410–415.

Schneider, K. R., Chang, A. and Goodrich, R. M. 2011. "Preventing Foodborne Illness: *Clostridium botulinum*." *University of Florida. Institute of Food and Agricultural Sciences (IFAS) Extension*.

Shimokomaki, M., E. Youssef, Y. and Terra, N. N. 2003. "CURING." In *Encyclopedia of Food Sciences and Nutrition (Second Edition)*, edited by Benjamin Caballero, 1702-1708. Oxford: Academic Press.

Sivamaruthi, B. S., Kesika, P. and Chaiyasut, C. 2019. "Toxins in fermented foods: prevalence and preventions—a mini review." *Toxins* 11 (1):4.

Škaljac, S., Petrović, L., Tasić, T., Ikonić, P., Jokanović, M., Tomović, V., Džinić, N., Šojić, B., Tjapkin, A. and Škrbić, B. 2014. "Influence of smoking in traditional and industrial conditions on polycyclic aromatic hydrocarbons content in dry fermented sausages (Petrovská klobása) from Serbia." *Food Control* 40:12-18.

Stanner, S., and H. Gibson-Moore. 2013. "Pregnancy: Safe Diets." In *Encyclopedia of Human Nutrition* (Third Edition), edited by Benjamin Caballero, 90-98. Waltham: Academic Press.

Takeda, S., Matsufuji, H., Nakade, K., Takenoyama, S., Ahhmed, A., Sakata, R., Kawahara, S. and Muguruma, M. 2017. Investigation of lactic acid bacterial strains for meat fermentation and the product's antioxidant and angiotensin-I-converting-enzyme inhibitory activities. *Animal Science Journal*, 88: 507–516.

Talon, R., Leroy, S. and Lebert, I. 2007. Microbial ecosystems of traditional fermented meat products: The importance of indigenous starters. *Meat Science*, 77: 55–62.

Tanasupawat, S., Phoottosavako, M. and Keeratipibul, S. 2015. "Characterization and lipolytic activity of lactic acid bacteria isolated from Thai fermented meat." *J App Pharm Sci* 5 (03):006-012.

ten Brink, B., Damink, C., Joosten, H. M. L. J. and In't Veld, J.H. 1990. "Occurrence and formation of biologically active amines in foods." *International journal of food microbiology* 11 (1):73-84.

Tirmenstein, M. A. and Mangipudy, R. 2014. "Aflatoxin." In *Encyclopedia of Toxicology* (Third Edition), edited by Philip Wexler, 104-106. Oxford: Academic Press.

Toldrá, F. and Reig, M. 2015. "The Biochemistry of Meat and Fat." In: F. Toldrá, Y. H. Hui, I. Astiasarán, J. G. Sebranek, and R. Talon. (Eds.), *Handbook of Fermented Meat and Poultry*, pp. 49–54. Wiley-Blackwell, Ames, Iowa, USA.

Toldrá, F., 2007. *Handbook of Fermented Meat and Poultry*. Blackwell Publishing, Oxford, UK.

Toldrá, F., Reig, M., Aristoy, M. C. and Mora, L. 2018. Generation of bioactive peptides during food processing. *Food Chemistry*, 267: 395–404.

Tosukhowong, A., Visessanguan, W., Pumpuang, L., Tepkasikul, P., Panya, A. and Valyasevi, R. 2011. "Biogenic amine formation in Nham, a Thai fermented sausage, and the reduction by commercial starter culture, *Lactobacillus plantarum* BCC 9546." *Food Chemistry* 129 (3):846-853.

Vaštag, Z., Popović, L., Popović, S., Petrović, L. and Peričin, D. 2010. Antioxidant and angiotensin-I converting enzyme inhibitory activity in the water-soluble protein extract from *Petrovac Sausage* (Petrovská Kolbása). *Food Control*, 21: 1298–1302.

Visessanguan, W., Benjakul, S., Riebroy, S. and Thepkasikul, P. 2004. Changes in composition and functional properties of proteins and their contributions to Nham characteristics. *Meat Science*, 66: 579–588.

Vulić, A., Vahčić, N., Hengl, B., Gross-Bošković, A., Jurković, M., Kudumija, N. and Pleadin, J. 2016. "Assessment of possible human exposure to ochratoxin A in Croatia due to the consumption of dry-cured and fermented meat products." *Food Additives & Contaminants: Part A* 33 (9):1428-1434.

Wanangkarn, A, Tan, F. J., Phumthong, N. and Boonsema, P. 2020. "Identification of lactic acid bacteria isolated from Thai fermented sausage for nitrite degradation ability." *Journal of Agricultural Technology* 16 (3):761-770.

Wanangkarn, A., Liu, D. -C., Swetwiwathana, A., Jindaprasert, A., Phraephaisarn, C., Chumnqoen, W. and Tan, F. -J. 2014. Lactic acid bacterial population dynamics during fermentation and storage of Thai fermented sausage according to restriction fragment length polymorphism analysis. *International Journal of Food Microbiology*, 186: 61–67.

Wang, X., Zhang, Y., Ren, H. and Zhana, Y. 2018. Comparison of bacterial diversity profiles and microbial safety assessment of salami, Chinese dry-cured sausage and Chinese smoked-cured sausage by high-throughput sequencing. *LWT - Food Science and Technology*, 90: 108–115.

Wilson, K. A., R. W. Kung, and S. D. Wetmore. 2016. "Chapter Seven - Toxicology of DNA Adducts Formed Upon Human Exposure to Carcinogens: Insights Gained From Molecular Modeling." In *Advances in Molecular Toxicology*, edited by J. C. Fishbein and J. M. Heilman, 293-360. Elsevier.

Wójciak, K. M., Libera, J., Stasiak, D. M. and Kołożyn-Krajewska, D. 2017. Technological aspect of *Lactobacillus acidophilus* Bauer, *Bifidobacterium animalis* BB-12 and *Lactobacillus rhamnosus* LOCK900 use in dry-fermented pork neck and sausage. *Journal of Food Processing and Preservation*, 41: e12965.

Wu, B. 2020. "Chapter 6 - Human health hazards of wastewater." In *High-Risk Pollutants in Wastewater*, edited by Hongqiang Ren and Xuxiang Zhang, 125-139. Elsevier.

Yadav, A. S. 2017. Antioxidant and antimicrobial profile of chicken sausages prepared after fermentation of minced chicken meat with *Lactobacillus plantarum* and with additional dextrose and starch. *LWT - Food Science and Technology*, 77: 249–258.

Zhang, Y., Hu, P., Lou, L., Zhan, J., Fan, M., Li, D. and Liao, Q. 2017. Antioxidant Activities of Lactic Acid Bacteria for Quality Improvement of Fermented Sausage. *Journal of Food Science*, 82: 2960–2967.

Zommiti, M., Bouffartigues, E., Maillot, O., Barreau, M., Szunerits, S., Sebei, K., Feuilloley, M., Connil, N. and Ferchichi, M. 2018. In vitro Assessment of the probiotic properties and bacteriocinogenic potential of *Pediococcus pentosaceus* MZF16 isolated from artisanal Tunisian meat "Dried Ossban." *Frontiers in Microbiology*, 9: 2607.

In: Fermented Foods
Editor: Oliver Kovalyov
ISBN: 978-1-53618-656-7

Chapter 5

FERMENTED SHRIMP: COMPOSITION, NUTRITIVE VALUE, MICROBES AND BENEFITS

Rungsima Daroonpunt[1,*], Engkarat Kingkaew[2] and Somboon Tanasupawat[2,†]

[1]Nutritional Therapy and Dietetics Division,
Faculty of Allied Health Sciences, Burapha University,
Chonburi, Thailand

[2]Department of Biochemistry and Microbiology,
Faculty of Pharmaceutical Sciences, Chulalongkorn University,
Bangkok, Thailand

ABSTRACT

Fermented shrimp products are traditional fermented foods produced in Southeast Asia (Thailand, Philippines, Vietnam, Indonesia, Myanmar, and Malaysia). *Ka-pi* is a traditional Thai fermented shrimp paste with

* Corresponding Author's Email: rungsima.da@go.buu.ac.th.
† Corresponding Author's Email: Somboon.T@chula.ac.th.

pinkish or purplish gray to a dark grayish brown-colored, strong odor and salty. On the other hands, *kung-chom* and *kung-som* are other kinds of Thai fermented shrimp paste with orange or red pinkish, sour and salty. Halophilic and lactic acid bacteria predominantly found in fermented shrimp products and play an important role in fermentation. Halophilic and related bacteria consist of various genera: *Alkalibacterium, Alkalibacillus, Aeromonas, Alcaligenes, Amphibacillus, Bacillus, Haloarcula, Halorubium, Jeotgalicoccus, Lentibacillus, Macrococcus, Micrococcus, Oceanobacillus, Pseudomonas, Salinicoccus, Salinivibrio, Staphylococcus, Sulfidobacillus, Vibrio* and *Virgibacillus*, while lactic acid bacteria composed of diverse genera: *Enterococcus, Lactobacillus, Lactococcus, Leuconostoc, Pediococcus, Sporolactobacillus, Tetragenococcus,* and *Weisella.* During fermentation, the protein and fats hydrolysis of shrimp is mediated by the action of microbial enzymes and endogenous enzymes derived from shrimp yield short-chain peptides, free amino acids and short-chain fatty acids which enhance the unique flavor and umami taste and contain biological activities. Additionally, the fermented microorganisms synthesize several metabolites (e.g., alcohol, aldehydes, esters, ketones, organic acids, and others) that contribute to the flavor, aroma, and other beneficially biological activities of fermented shrimp products. Hence, fermented shrimps can be highly nutritious food because of its composition and high digestibility that contribute to health. This chapter display and discuss the production, composition, microorganisms, health benefits, and health risks of fermented shrimp products.

1. INTRODUCTION

Fermented fish products are the popularly fermented foods and mostly consumed in the Southeast Asian countries (Chaves-López et al. 2014, Irianto 2017). There are various kind of fermented fish products available, include watery products such as fish sauce (*nam-pla*) in Thailand, *patis* in the Philippines, *nouc-mam* in Vietnam, *kecap ikan* or *bakasang* in Indonesia, oyster sauce, *hoi-sin* sauce, and paste products such as fish and shrimp pastes also known as *ka-pi, kung-chom* and *kung-som* in Thailand (Rattagool 1985, Ishige 1993, Phithakpol and Kasetsat 1995). Though these products are known with various names, their main role is the same to provide a salty and umami taste (Hajeb and Jinap 2012). Fermented shrimp products are a trendy Thai fermented fish produced according to family or local geographical preferences, especially in the central, eastern, southern, northern, and northeastern parts of Thailand (JIRCAS 2011a, b, c).

Fermented shrimp products (*Ka-pi, Kung-chom*, and *Kung-som*) are traditional foods in the Southeast Asian countries since the fifteenth century and consumed as staples, side dishes, or condiments/seasonings in daily foods (Pongsetkul 2018, JIRCAS 2011a, b). The major ingredients for fermented shrimp production are shrimps of the genera *Acetes*, *Lucifer*, *Mesopodopsis*, and *Mysids*. The genus *Acetes* is the most ordinary raw materials to produce shrimp sauce, paste, and other fermented products such as *ka-pi* (Thailand) which had high glutamic acid and lysine contents (Kleekayai, Saetae, et al. 2015, Hajeb and Jinap 2012). Noticeably, fermented shrimp products are plentiful sources of peptides, amino acid, short-chain fatty acids, minerals, probiotics and also bioactive compounds (Dangkhaw, Maneerat, and Sumpavapol 2012, Salampessy, Kailasapathy, and Thapa 2010). During fermentation, raw materials are converted into products through activities of endogenous enzymes or microorganisms (Pongsetkul et al. 2014). Although some products have a similar process, some ingredients used can be varied, leading to different characteristics, especially color, flavor, and taste. Fermented shrimp paste prepared from the mixing of small shrimp with salt and fermentation which is widely used as condiments or an essential ingredient in many cuisines due to its flavor and nutritional properties .(Hajeb and Jinap 2012) In addition, these fermented products will make the food more flavorful and appetizing. Thus, it has become an essential ingredient in curries and dipping sauces for a more appealing taste with unique textures, flavors, and aromas. (Leong et al. 2009)

2. Shrimp Paste (*Ka-Pi*)

Shrimp paste, a pinkish or purplish gray to a dark grayish brown with strong odor, salty, and umami taste and paste-like consistency is extensively consumed as condiments/seasoning ingredient in Southeast

Asian countries. There are known by various names in different countries; for example, *ka-pi* (Thailand), *bagoong* (Philippines), *shiokara* (Japan), *saewoojeo* (Korea*), mam ruoc* (Vietnam), *terasi* (Indonesia), *ngapi* (Myanmar) and *belacan* (Malaysia and Brunei) (Wittanalai, Rakariyatham, and Deming 2011, Chaijan and Panpipat 2012). *Ka-pi* is generally prepared from the small shrimp or krill such as *Acetes indicus*, *Acetes japonicas*, *Acetes erythraeus* (*Koei Taa Deang*, red paste product), *Mesopodopsis orientalis* (*Koei Taa Dam*, black paste product), and *Acetes vulgaris* (*Koei*), mixed with salt and sun-dried to decrease the moisture content and then the fermented for at least 3 months until the desired aroma has developed (Faithong et al. 2010, Kleekayai, Saetae, et al. 2015). The major processing of this product is salting, fermenting, and aging. During fermentation, the protein hydrolysis of shrimp paste is mediated by the action of endogenous enzymes derived from shrimp and microbial enzymes yield short-chain peptides and free amino acids which enhance the unique flavor and umami taste. The fermentation process is influenced by several variables including the microflora present in the raw material, the proteolytic activity of raw material and microorganism, presence or absence of oxygen, nutritional value of the raw material, pH, temperature, the duration of the fermentation process, concentration of NaCl, condition of the fermentation process and the presence of visceral enzymes (Phithakpol and Kasetsat 1995).

2.1. Production of Shrimp Paste (*Ka-pi*)

The shrimp paste production in each region was different in terms of raw material, the ratio of salt used, fermentation process, and time. *Ka-pi* is generally prepared by the small shrimp or krill. It depends on the types of shrimps available in each country and then mixed with salt at a ratio of 3-10:1 (w/w) and sun-dried for 3 days to decrease the moisture content and then ground into a fine paste before being compacted in a container. The paste is allowed to ferment at 1-3 months at ambient temperature (28-30 °C) until the desired aroma has developed (Faithong et al. 2010,

Pongsetkul et al. 2014, Prapasuwannakul and Suwannahong 2015). This procedure is shown in Figure 1. For the *ka-pi* production, salt plays a key role in preservation by inhibiting the growth of spoilage and pathogenic microorganisms and also led to the lower moisture content and water activity (A_w) of shrimp paste products and help to extend the shelf-life of this product.

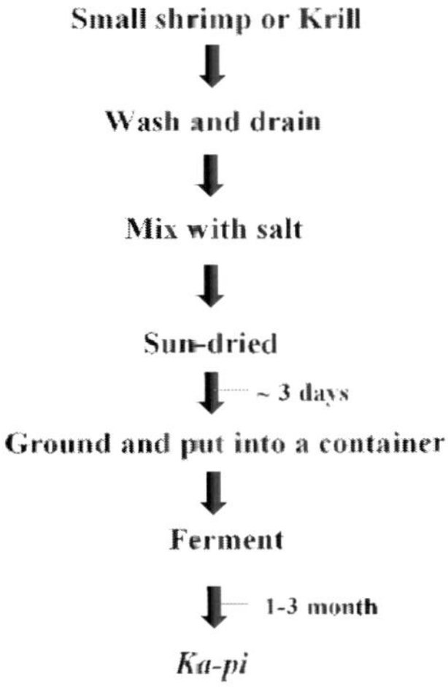

Figure 1. *Ka-pi* production.

2.2. Composition of *Ka-pi*

This product contained the percentage of moisture content (%) and water activity (A_w) between 23.0 to 52.54 and 0.64 to 0.74, respectively, that was classified as an intermediate moisture food. The low A_w of shrimp paste can preserve the food from microbial spoilage, limit the growth of food pathogens, and could be associated with the prolonged shelf-life of shrimp paste products. The pH of the shrimp paste was shown between 5.5 to 8.4. The slightly basic pH might be affected by the degradation of products liberated during fermentation or the formation of volatile basic compounds such as ammonia (Prapasuwannakul and Suwannahong 2015).

When the hydrolysis proceeded, the small proteins or peptides might undergo decomposition and this could favor the formation of low-molecular-weight basic compounds (Pongsetkul et al. 2014). The shrimp paste products had different colors, *L**(lightness), *a**(redness) and *b**(yellowness) value ranged from 29.6-51.84, 5.24-15.39, and 4.56-17.91, respectively. It might be caused by the difference pigment contents in raw materials, process as well as ingredients added. Browning intensity most likely caused by the formation of Maillard reaction products (MRPs) were concomitantly observed throughout fermentation, as evidenced by the decreases in lightness (*L**-value), but the increases in redness (a*-value) and yellowness (b*-value) (Faithong and Benjakul 2014). The shrimp paste can be seen as a good source of minerals such as Ca and Fe (Pilapil et al. 2015), while the Na content is quite high owing to the large amount of salt used in the production process. The chemical compositions of shrimp paste were shown in Table 1.

2.3. Nutritive Value of *Ka-pi*

In *ka-pi*, the predominant amino acids were glutamic acid (70.1-593.9 μg/g), lysine (112.7-546.3 μg/g), leucine (29.5-544.9 μg/g), isoleucine (18.5-364.3 μg/g) and aspartic acid (96.5- 281.7 μg/g). In addition, Gamma-aminobutyric acid (GABA) was detected (0.7-2.6 μg/g) (Daroonpunt, Uchino, et al. 2016). GABA has several well-known physiological functions and positive effects on many metabolic disorders such as anti-anxiety and anti-depressant considered a potent bioactive compound (Diana, Quilez, and Rafecas 2014). The concentration of free glutamic acid in Malaysian *belacan*, was reported to be 180.0 – 4,207.0 mg/100 g (Jinap et al. 2010), 1,550.0 mg/100 g (Yoshida 1998). The essential amino acid (EAA) of *terasi* in Indonesia was dominated by lysine (21.26 - 42.52 mg/g), followed by valine (20.56 - 41.11 mg/g) and leucine (16.20 - 32.40 mg/g), while non-essential amino acid (NEAA) was dominated by glutamic acid (28.94-57.88 mg/g), asparagine (1.84 - 3.68 mg/g) and alanine (11.77 - 23.54 mg/g) (Ali et al. 2020). The most abundant amino acid of the Korean and Bruneian shrimp pastes was glutamic acid, followed by aspartic acid and leucine (Kim et al. 2014). The high amounts of glutamic acid in shrimp paste were indicated the rich of umami taste. Kim et al. (2014) reported that the saturated fatty acid, monounsaturated fatty acid and polyunsaturated fatty acid content of Brunei's *belacan* and Korean's *Saewoojeot* shrimp paste were 31.16 and 23.74%, 21.63 and 18.99%, and 47.21 and 57.27%, respectively. The traditional and commercial shrimp pastes in the Philippines varied in their content of saturated fatty acids (32.6-39.1g/100g fatty acid methyl esters (FAME)), monounsaturated fatty acids (15.1-18.7 g/100g FAME) and polyunsaturated fatty acids (30.7-40.8 g/100 g FAME) (Pilapil et al. 2015). The predominant fatty acids (C16:0, C18:0, C18:1n9t, C20:1n7c, C20:5n3, and C22:6n3) and amino acids (aspartate, glutamate, alanine, leucine, lysine, arginine, and proline) in shrimp paste were reported by Cai et al. (2017). Lipid oxidation occurred during *ka-pi* production, probably owing to the high content of polyunsaturated fatty acids. These lipid oxidation products could affect odor and flavor characteristics of *ka-pi*) Pongsetkul et al. (2017.

The volatile compounds had been detected in shrimp paste including aldehydes, alcohol compounds, carbonyl, acids, esters, ketones, hydrocarbons, N-containing, S-containing, and phenolic compounds (Ali et al. 2020, Lv et al. 2020) and the typical volatile compounds contributing to the flavor of shrimp paste were found as follows: dimethyl disulfide, dimethyl tetrasulfide, dimethyl trisulfide, 2,3,5-trimethyl-6-ethyl pyrazine, ethyl-2,5-dimethyl-pyrazine, phenol and indole (Fan et al. 2017). According to Aryanta (2000), the deamination and decarboxylation of amino acids to form lower fatty acids and amides, producing the characteristic flavor of *terasi*. The nitrogen-containing compounds such as pyrazines comprised the major of volatile compounds that were predominant in fermented shrimp paste and they are probably important in the characteristic flavor of shrimp paste (Cai et al. 2017, Cha and Cadwallader 1995).

Table 1. Chemical composition of shrimp paste (*ka-pi*) and related products

Product	Moisture (%)	A_w	Protein (%)	Fat (%)	pH	Ash (%)	NaCl (%)
Ka-pi[a]	33.95-52.19	0.64-0.72	17.9-42.8	ND	6.90-8.10	ND	7.0-10.85
Ka-pi[b]	37.36-46.85	0.70-0.74	18.95-25.14	0.69-2.05	7.01-7.71	0.69-2.05	19.78-22.96
Ka-pi[c]	33.79-52.54	0.70-0.77	29.44-53.27	1.41-3.67	7.01-8.40	33.80-50.50	22.77-35.47
Ka-pi[d]	36.78-49.93	ND	20.14-25.12	1.53-2.22	7.44-7.66	24.13-28.99	19.91-24.73
Ka-pi[e]	37.1-51.4	ND	7.7–24.0	0.4-32.7	5.5-6.4	ND	13.6-21.2
Belacan[f]	23.0-47.4	ND	ND	ND	6.8-7.6	ND	17.5-39.7
Bruneian *belacan*[g] (mean)	47.92	0.728	30.38	0.63	7.56	19.15	14.94
Philippines shrimp paste[h]	ND	0.70-0.74	12.9-15.3	0.50-1.94	6.8-7.7	ND	4.04-5.15
Saewoojeot[g] (mean)	26.96	0.682	21.70	4.89	7.68	45.83	13.46
Terasi[i] (mean)	37.41	ND	25.42	6.11	7.53	29.12	16.75

[a]Samples from Songkhla and Nakhon Sri Thammarat provinces (Daroonpunt, Uchino, et al. 2016). [b]Planktonous shrimp from Klongkone district, Samut Songkram province (Prapasuwannakul and Suwannahong 2015). [c]Samples from Songkhla, Ranong, Krabi, Sa-tun, Samut Sakhon, Rayong, Chachoengsao and Samut Songkram provinces (Pongsetkul et al. 2014). [d]Planktonous shrimp from Songkhla and Trang provinces (Faithong et al. 2010). [e]Sample sources were not reported (Phithakpol 1993). [f]Padilah et al. (2018); [g]Kim et al. (2014); [h]Pilapil et al. (2015); [i]Surono and Hosono (1994a). ND, no data.

3. *KUNG-CHOM*

Kung-chom, a traditional Thai fermented shrimp product that popularly consumed in the lower northern and northeastern regions of Thailand, is made by the mixing of small shrimp with salts or ground roasted rice and fermentation. It is orange or red pinkish semi-solid food with sour and salty taste, but aroma and color are relied on ingredients. The fermentation of *kung-chom* is conducted by LAB and some halophilic bacteria that produce several beneficial metabolites, digest complex nutrient to easily absorbed molecules, develop food characteristics, and also preserve and prolong shelf-life of product (Pakawadee 2016, JIRCAS 2011a, Prachasitthisak et al. 2008).

3.1. *Kung-chom* Production

Kung-chom production is naturally fermentation by shrimp normal flora. The whole small shrimp is washed and drained. Next, 10 parts of *Macrobrachium lanchesteri* (shrimp) gently mixed with 1 part of salt, 3 parts of ground roasted rice and 0.4 parts of minced garlic (w/w). Thereafter, the mixtures are packed in closed container and incubate at ambient temperature for 3-5 days. However, some proportion of each ingredient is various and may presently use fish sauce instead of salt. After incubation time, it can be conventionally consumed with cooked rice, mixing spices as salad, and using as condiment for dipping vegetables. Normally, the shelf-life of *kung-chom* is about 2 weeks. This procedure is briefed and displayed in Figure 2 (JIRCAS 2011a).

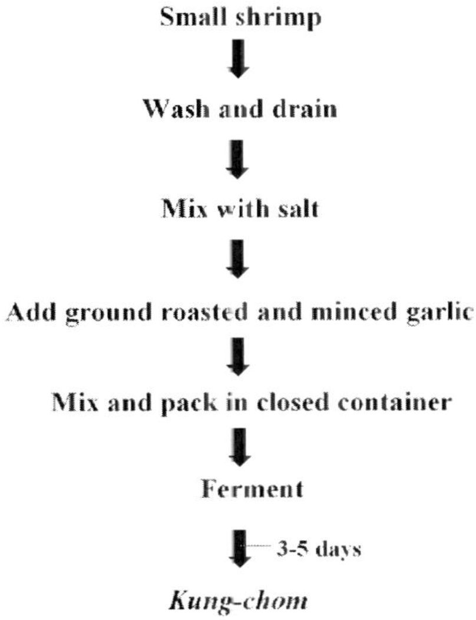

Figure 2. *Kung-chom* production (JIRCAS 2011a).

4. *KUNG-SOM*

Furthermore, *kung-som,* a traditional Thai fermented shrimp products, is widely consumed in the southern and northeastern part of Thailand. It is made by mixing small shrimp with salt, sugars, minced cook rice and water. After that, the mixtures are incubated in anaerobic condition in anaerobic containers at ambient temperature. Importantly, some ingredients of *kung-som* are different from each region, for example the using Jaggery palm sugar instead sugar, supplementation of washing rice water or ground roasted rice, and the replacing salts with fermented shrimp paste (*ka-pi*). The bacterial community of *kung-som* is diverse and complex, but the main and minor microorganisms of *kung-som* were lactic acid bacteria (LAB) and halophilic bacteria, respectively. These bacteria play role in fermentation processes to develop the good sensory characteristics and improve digestibility and nutritive value (TCPS324 2004). For the *kung-som* properties, it is solid with sour and salty taste, and has orange pinkish color (JIRCAS 2011b, Saelao et al. 2016, Sanchart et al. 2015, Thongrucka, Saelaoa, and Sumpavapolb 2017).

4.1. *Kung-som* Production

Kung-som production, the whole shrimp (*Macrobrachium lanchesteri*) is firstly washed and drained. Secondly, shrimp is mixed with salt in proportion of 8:1 by weight and then left overnight at ambient temperature. Fourthly, cooked rice and minced garlic are added in the ratio 20:4:1 by weight and mixed well. Finally, the product is packed in a closed container and incubated at ambient temperature for 5-7 days. At the end of the incubation period, it can be conventionally consumed with cooked rice or mixed with spices. Normally, the storage time of *kung-som* is about 3 weeks after opening (JIRCAS 2011b, Nantida 2018). This procedure is summarized and shown in Figure 3.

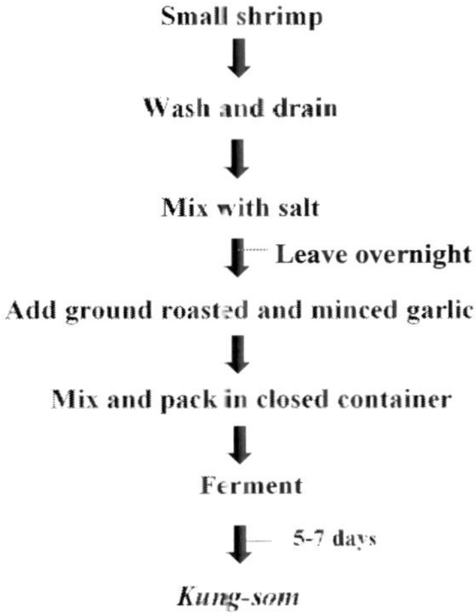

Figure 3. *Kung-som* production (JIRCAS 2011b).

4.2. Composition of *Kung-chom* and *Kung-som*

Chemical composition of *kung-chom* and *kung-som* is depended on shrimp species, size, feeds, farm environments, spawning period, production and fermentation process. For instance, One hundred grams of *kung-chom* supplemented with 5% (w/w) of NaCl has high crude proteins (Pakawadee 2016, Sikorski, Kołakowska, and Pan 1990). From the chemical composition, it can be possibly summarized that *kung-chom* and *kung-som* are classified as low-salt food product that contain salt (NaCl) about 3-9% (Martínez-Álvarez et al. 2017). Moreover, the study of Faithong et al. (2010) and report of JIRCAS (2011b) stated that *kung-som* has a moisture content (66.3–83.71%) with protein content (4.53–11.4%) and salt content (2.6–5.9%), while carbohydrate contents ranged from 6.93% to 15.75%. Sugar concentrate supplemented for LAB fermentation is the main source of carbohydrate. Thus, *kung-som* has acidic pH, 3.62–5.6, which was mainly due to lactic acid generated during fermentation. The proximate chemical compositions of *kung-chom* and *kung-som* were shown in Table 2.

Table 2. Chemical composition of *Kung-chom* and *Kung-som*

Product	Moisture (%)	A_w	Protein (%)	Fat (%)	pH	Ash (%)	NaCl (%)	Fibre (%)	Total Acidity (%)
Kung-chom[a]	60.7 - 73.2	0.85	6.8 - 14.2	0.9 - 2.8	3.9 - 5.0	3.2 - 9.4	3.2 - 9.4	Trace - 1.8	1.1 - 3.8
Kung-chom[b]	ND	ND	12.10 ± 1.47	ND	4.86 ± 0.23	ND	8.89 ± 0.96	ND	1.47 ± 0.23
Kung-chom[d]	75.62	ND	6.62	0.54	3.96 – 4.69	6.26	ND	ND	2.45 - 2.72
Kung-som[c]	73.35 ± 0.15	ND	6.38 ± 0.09	0.11 ± 0.01	3.89 ± 0.02	4.41 ± 0.18	2.81 ± 0.04	ND	ND
Kung-som[c]	80.71 ± 0.09	ND	7.21 ± 0.03	0.15 ± 0.04	3.71 ± 0.01	5.00 ± 0.03	3.74 ± 0.00	ND	ND
Kung-som[d]	78.65	ND	4.53	0.12	3.62 - 4.03	4.90	ND	ND	2.30 - 3.57
Kung-som[e]	66.3 - 83.1	0.85- 0.87	8.3 - 11.4	0.0 - 2.1	3.7 - 5.6	3.7 - 7.4	2.6 - 5.9	0.4 - 2.8	2.7 - 4.4

[a]Sample sources were not reported (JIRCAS 2011a). [b]Samples from Buriram province (Prachasitthisak et al. 2008). [c]Samples from Songkhla and Trang provinces (Faithong et al. 2010). [d]Samples from Bangkok (Chaisri et al. 2005). [e]Sample sources were not reported (JIRCAS 2011b). Mean ± SD from triplicate determinations. Acidity (%) as lactic acid. ND; no data.

4.3. Nutritive Value of *Kung-som* and *Kung-chom*

Fermented shrimp products are nutritious food, and sources of protein and minerals through fermentation and show beneficial biological activities (Peralta et al. 2008, Montano, Gavino, and Gavino 2001, Chuon et al. 2014, Pakdeeto, Tanasupawat, Thawai, Moonmangmee, Kud, et al. 2007, Kobayashi et al. 2003, Faithong and Benjakul 2014, Faithong et al. 2010, Namwong et al. 2009). One hundred grams of *kung-chom* supplemented with 5% (w/w) of NaCl consists of moisture (61.20%), ash (6.10%), protein (12.40%) and carbohydrate (19.10%), NaCl (3.78%), but just contains 136.80 kcal/100g. From the chemical composition (Table 2), it is possibly recapitulated that *kung-chom* and *kung-som* are classified as high-moisture and low-fat food products (Pakawadee 2016). Moreover, the fermented shrimps are a great source of polyunsaturated fatty acids (PUFA) such as docosahexaenoic acid (DHA) and eicosapentaenoic acid

(EPA), which are vital for the growth and functional development of the brain (memory, cognition and learning), and also required for brain functional maintenance. In addition, the DHA and EPA play role in antitumor activity, regulation of lipid and glucose metabolism, anti-inflammatory, protection against nephrotoxicity and improvement on exercise performance (Cai et al. 2017, Horrocks and Yeo 1999, Hyne et al. 2009, Montano, Gavino, and Gavino 2001, Zhang et al. 2019).

5. MICROBIOLOGY OF FERMENTED SHRIMP PRODUCTS

The microbial diversity of Thai traditional and commercial fermented shrimp products is very numerous and is important for understanding the role of microorganism in fermentation process to improve a better flavor and quality of the products. In *ka-pi*, the total bacterial cell count ranged from $1.3x10^3$ - $2.9x10^5$ CFU/g was reported (Daroonpunt, Uchino, et al. 2016), while *terasi* had total viable bacterial counts of $4x10^5$ CFU/g and a halophilic count of $1.1x10^5$ CFU/g (Surono and Hosono 1994a). While *kung-som,* had total viable bacterial counts of 10^5-10^8 CFU/g and a LAB count of $>10^7$ CFU/g (Saelao et al. 2016). In *kung-chom,* Prachasitthisak et al. (2008) reported that the total bacterial cell count ranged from 10^4-10^8 CFU/g.

From the previous studies, it can be concluded that microorganisms in fermented shrimp are belonged to two major groups as halophilic bacteria and lactic acid bacteria (LAB). Numerous halophilic and related bacteria have been found in *ka-pi* and *terasi* such as *Micrococcus* (Surono and Hosono 1994b), *Pseudomonas* (Surono and Hosono 1994a, Murwani et al. 2015), *Kurthia* (Surono and Hosono 1994a, Murwani et al. 2015), *Lentibacillus* (Pakdeeto, Tanasupawat, et al. 2007a, Booncharoen et al. 2019), *Salinicoccus* (Pakdeeto, Tanasupawat, et al. 2007b), *Alkalibacterium* (Ishikawa et al. 2009), *Oceanobacillus* (Namwong et al. 2009), *Vibrio, Caryophanon*, *Sulfidobacillus*, *Aeromonas, Alkaligenes*, *Amphibacillus* (Murwani et al. 2015), *Bacillus* (Surono and Hosono 1994a,

Surono and Hosono 1994b, Kaewklom et al. 2013, Sudaryati and Rahmansyah 2014, Murwani et al. 2015, Pongsetkul et al. 2018, Daroonpunt, Saeng-in, and Tanasupawat 2019, Daroonpunt et al. 2019), *Virgibacillus* (Daroonpunt, Tanasupawat, et al. 2016, Kumaunang et al. 2019), *Staphylococcus* (Daroonpunt, Saeng-in, and Tanasupawat 2019), *Jeotgalicoccus* and *Macrococcus* (Phewpan et al. 2020). Additionally, LAB has been discovered in *ka-pi* and *terasi* such as *Tetragenococcus* (Kobayashi et al. 2003, Tanasupawat 2009), *Lactobacillus* (Amalia and Agustini 2018), *Bacillus* and *Sporolactobacillus* (Surono and Hosono 1994a). Thus, several studies have reported the distribution of bacteria in *ka-pi* and *terasi* as shown in Table 3.

For *kung-chom* and *kung-som*, several studies stated that many microorganisms are mostly belonged to the group of LAB, which includes *Enterococcus* (Sangjindavong et al. 2006), *Tetragenococcus* (Tanasupawat 2009), *Lactobacillus* (Tanasupawat 2009, Hwanhlem et al. 2010, Sanchart et al. 2017), *Lactococcus* (Rattanachaikunsopon and Phumkhachorn 2008, Saelao et al. 2017). While, halophilic and related bacteria are minorly found in them. They include *Bacillus* (Phoottosavako et al. 2015), *Staphylococcus* (Tanasupawat et al. 1992). Nevertheless, some researchers reported the microorganisms in fermented shrimp by using conventional microbiological method with selective culture media, which did not represent the entire microorganism diversity (Ben Omar and Ampe 2000). Thus, several recent studies have to develop and utilize polymerase chain reaction (PCR) techniques (i.e., PCR-Denaturing gradient gel electrophoresis (DGGE)) to determine the diversity and changes of the microbes in food due to more information providing and widely used for detection, monitoring, classification and identification of bacterial diversity (Ercolini 2004, Cocolin et al. 2001). The studies of Saelao et al. (2016), Sanchart et al. (2015) and Thongrucka, Saelaoa, and Sumpavapolb (2017) used PCR-Denaturing gradient gel electrophoresis (DGGE) technique and reported the microbial distribution in *kung-som* include the strains in genera *Bacillus* (Saelao et al. 2016), *Staphylococcus* (Saelao et al. 2016,

Table 3. Distribution of bacteria in fermented shrimp products (*Ka-pi* and *Terasi*)

Fermented products	Halophilic and related bacteria	References	Lactic acid bacteria	References
Ka-pi	*Lentibacillus kapialis*	Pakdeeto, Tanasupawat, et al. (2007a)	*Tetragenococcus halophilus, T. muriaticus*	Tanasupawat (2009)
	Salinicoccus siamensis	Pakdeeto, Tanasupawat, et al. (2007b)		
	Alkalibacterium kapii	Ishikawa et al. (2009)		
	Oceanobacillus kapialis	Namwong et al. (2009)		
	Bacillus amyloliquefaciens	Kaewklom et al. (2013)		
	Virgibacillus kapii	Daroonpunt, Tanasupawat, et al. (2016)		
	B. subtillis	Pongsetkul et al. (2018)		
	Lent. lipolyticus	Booncharoen et al. (2019)		
	B. tequilensis, B. infantis B. flexus, B. paramycoides, and *Staphylococcus cohnii* subsp. *urealyticus*	Daroonpunt, Saeng-in, and Tanasupawat (2019)		
	B. salacetis	Daroonpunt et al. (2019)		
	Vir. halodenitrificans	Kumaunang et al. (2019)		
	Jeotgalicoccus psychrophilus, Macrococcus caseolyticus	Phewpan et al. (2020)		
Terasi	*Pseudomonas aeruginosa, P. cepacian, P. alcaligenes, P. fluorescens, Kurthia gibsonii, B. brevis, B. sphaericus, B. subtilis, B. circulans,* and *B. polymyxa*	Surono and Hosono (1994a)	*Sporolactobacillus inulinus, B. coagulans*	Surono and Hosono (1994a)
			T. halophilus, T. muriaticus	Kobayashi et al. (2003)
			Lactobacillus plantarum	Amalia and Agustini (2018)
	B. pumilus, B. megaterium, and *Micrococcus kristinae*	Surono and Hosono (1994b)		
	B. subtilis	Sudaryati and Rahmansyah (2014)		

Table 4. Distribution of bacteria in fermented shrimp products (*Kung-chom* and *Kung-som*)

Fermented products	Halophilic and related bacteria	References	Lactic acid bacteria	References
Kung-chom	*Bacillus* sp. *Sta. piscifermentans*	Phoottosavako et al. (2015) Tanasupawat et al. (1992)	*Enterococcus faecalis* *Lb. pentosus, Lb. plantarum, Lb.farciminis,* and *T. halophilus* *Lactococcus lactis* subsp. *lactis*	Sangjindavong et al. (2006) Tanasupawat (2009) Rattanachaikunsopon and Phumkhachorn (2008)
Kung-som	*Bacillus* sp., *Sta. carnosus* *Salinivibrio sharmensis* *Sta. pasteuri,* *Mac. caseolyticus*	Saelao et al. (2016) Sanchart et al. (2015) Thongrucka, Saelaoa, and Sumpavapolb (2017)	*Lb. plantarum* *T. halophilus, Lb. farciminis, Lb. plantarum, Lb. acetotolerans, Lb. rapi, Lb. crustorum, Lb. fermentum, Lb. reuteri, Sta. piscifermentans, Weissella thailandensis*, and *W. cibaria* *Lb. farciminis, Lb. plantarum, Lc. garvieae, T. halophilus, Weissella thailandensis,* *Pediococcus pentosaceus* *Lc. lactis* *Lb. futsaii* *En. faecalis, En. sanguinicola, W. cibaria, W. confuse, Leuconostoc pseudomesenteroides, Ped. argentinicus, Lc. lactis* subsp. *lactis, Lc. garvieae, Lb. curvatus, Lb. sakei,* and *Lb. plantarum*	Hwanhlem et al. (2010) Sanchart et al. (2015) Saelao et al. (2016) Saelao et al. (2017) Sanchart et al. (2017) Thongrucka, Saelaoa, and Sumpavapolb (2017)

Thongrucka, Saelaoa, and Sumpavapolb 2017), *Macrococcus* (Thongrucka, Saelaoa, and Sumpavapolb 2017), *Salinivibrio* (Sanchart et al. 2015), *Weisella* (Sanchart et al. 2015, Saelao et al. 2016, Thongrucka, Saelaoa, and Sumpavapolb 2017), *Pediococcus,* and *Leuconostoc* (Saelao et al. 2016, Thongrucka, Saelaoa, and Sumpavapolb 2017). The distribution of bacteria in *kung-chom* and *kung-som* shown in Table 4.

Furthermore, the bacterial communities of Thai traditional fermented shrimp paste (*ka-pi)* were detected by next-generation sequencing (NGS) technique that showed the inhabitance of various halophilic bacteria includes the strains in genera *Lentibacillus*, *Salinococcus*, *Salimicrobium, Staphylococcus, Bacillus, Vagococcus, Halorubium, Salinibacter, Haloarcula, Alkalibacterium*, *Alkalibacillus, Micrococcus, Photobacterium, Salinivibrio,* and *Virgibacillus* (Phewpan et al. 2020).

From the several studies, it can be possibly concluded that diverse LAB and halophilic bacteria from different regions play important roles in fermentation. These microorganisms may relate to good flavor, aroma and texture of products, and improve digestibility and nutritive value. For example, *St. carnosus* exhibited nitrite and nitrate reductase activity, that enhance the flavor, promote the red color development and also help preservation by inhibition of the lipid oxidation and prevention rancidity. Importantly, the pathogenicity of *Sta. carnosus* has not yet been reported (Guan, Cho, and Lee 2011, Saelao et al. 2016). Halophilic and lactic acid bacteria (i.e., *Halobacterium* sp., *Halomonas* sp., *T. muriaticus*, *T. halophilus*) can produce some lipolytic enzymes or proteolytic enzymes. These microbial enzymes play a significant role in hydrolysis of lipids and proteins in shrimp throughout the process. Those degradation might contribute to the desired characteristics of final products, especially flavor or taste and they also exhibit antioxidant activities (Pongsetkul et al. 2017, Hinrichsen, Montel, and Talon 1994, Chuea-nongthon et al. 2017, Udomsil et al. 2011, Akolkar, Durai, and Desai 2010, Prapasuwannakul and Suwannahong 2015). The proteolytic and lipolytic bacteria counts increased during fermentation and reached the highest level at day 30 of fermentation (Pongsetkul et al. 2017). *T. halophilus* can reduce dimethyl disulfide, which is an undesirable compound contributing to a fecal note.

Lv et al. (2020) studied bacterial community succession and volatile compound changes during fermentation of shrimp paste from Chinese and indicated that the bacteria *Pseudoalteromonas*, *Staphylococcus*, and *Oceanobacillus* were important in the formation of volatile compounds. *Pseudoalteromonas* positively correlated with ketones, alcohols, esters, and acids; *Staphylococcus* positively correlated with aldehydes and acids; and *Oceanobacillus* positively correlated with aldehydes and pyrazines (Lv et al. 2020). In addition, *T. halophilus* strains improved volatile compounds and amino acid profiles, besides lowering biogenic amine composition (Udomsil et al. 2011). Lee, Jung, and Jeon (2015) stated that the presence and growth of *Tetragenococcus* sp. was concurrented with the increase of lactate concentration, which might be released from ingredients. Lactate acts as an energy source in skeletal muscle fibres, promotes energy sufficiency during exercise, and also supports a role for lactate in enhancing human oxidative capacity. Furthermore, lactate-astrocyte and lactate-neuron shuttles enable lactate to supply energy to support cognitive function (Todd 2014). *Lb. curvatus, Lb. plantarum* and *Lb. sakei* strains were the main acid producers in the *kung-som* fermentation (Thongrucka, Saelaoa, and Sumpavapolb 2017). These strains can tolerate to extremely acidic condition and also potentially produce lactic acid for the tangy flavor (Aymerich et al. 2003). *Lb. curvatus* strain resulted in a swift pH reducing, which affects the characteristic development of fermented foods (Visessanguan et al. 2006). Additionally, LAB, especially *Lactobacillus* and *Lactococcus* strains decrease the pH of the product that is one of the key factors to ensure quality and safety. Lactic acid, bacteriocin and bacteriocin-like compounds can inhibit pathogenic and food spoilage bacteria; consequently prevent from infections and food poisoning. *Lc. lactis* subsp. *lactis* TFF 211 isolated from *kung-chom* to produce bacteriocin (nisin). Strain TFF 221 exhibited antimicrobial activity against foodborne pathogen (Rattanachaikunsopon and Phumkhachorn 2008). Dangkhaw, Maneerat, and Sumpavapol (2012) reported that *En. hirae* K34, *Lb. plantarum* K39, and *Lb. plantarum* K50 isolated from *kung-som*, showed antimicrobial activity against six foodborne pathogens including *B. cereus*, *Escherichia coli*, *Listeria monocytogenes*, *Sta. aureus*, *Salmonella*

Typhimurium and *Vibrio cholera*. Strain D10SM20, D8SM21, D10SM16 and D6SM3 were isolated from *kung-som* and were identified as *Lb. plantarum*. Four strains showed antimicrobial activity against pathogenic bacteria (Hwanhlem et al. 2010). Saelao et al. (2017) reported that *Lactococcus lactis* KTH0-1S isolated from *kung-som* can produce bacteriocin that inhibits foodborne pathogens and food spoilage bacteria. Moreover, LAB of fermented shrimp are normally used as probiotics, live microorganisms that, when administered in appropriate amounts, confer a health benefit on the host" (Hill et al. 2014). Probiotics have been associated with a range of various health benefits such as improvement of intestinal health, cholesterol-lowering effect, anti-pathogenic activity, immunomodulation, reducing in fecal enzymes and mutagenicity, cancer prevention, and reduction of the risk of gastrointestinal disease in several studies (Ohashi and Ushida 2009, Kechagia et al. 2013). Therefore, the understanding of microbial diversity in fermented foods is important. The information of the microorganisms involving fermentation will be useful for starter selection, optimization and production, which could lead to a better utilizing local microbial resources to promote unique flavor quality of the product in each region.

6. Biological Activities and Health Benefits

Fermented shrimp products containing high concentration of sodium chloride may not claim to be functional foods, but these products can be identified as rich sources of bioactive amino acids and peptides. Numerous bioactive compounds may be found or produced in raw material and fermentation processes. Bioactive peptides are short fragments of proteins with 2-20 amino acid residues and the sequences and amino acid composition of peptides can affect the activity of biopeptides. These compounds have been associated with the beneficial to promote health status due to these compounds contain various biological activities. This part shows fermented shrimp products with some biological activities (i.e., antioxidant, antihypertensive, anticoagulant and fibrinolytic activities) (Ichimura et al. 2003, Faithong et al. 2010, Wong and Mine 2004, Broncano et al. 2012, Shahidi and Zhong 2008, Pihlanto-Leppälä, Rokka, and Korhonen 1998).

6.1. Antioxidant Activity

Reactive oxygen species (ROS), free radicals, and reactive nitrogen species (RNS) are ordinarily generated during metabolism. These species play a dual role as both toxic and beneficial compounds. For low or moderate levels, ROS and RNS provide beneficial effects on cellular responses and immune function. However, when an overload of free radicals, their accumulation in the body generates a condition named oxidative stress. Oxidative stress plays a major part in the development of chronic and degenerative illness such as cancer, autoimmune disorders, aging, cataract, rheumatoid arthritis, cardiovascular and neurodegenerative diseases. The result of massive free radicals are the body's ability devastation and adversely biomolecules modification (i.e., proteins, DNA, and membrane lipids) (Pham-Huy, He, and Pham-Huy 2008, Lobo et al. 2010, Najafian and Babji 2015). Therefore, a balance between free radicals generated in the body and antioxidants is importantly necessary. Exogenous and/or endogenous antioxidants act as free radical scavengers that can delay and against oxidation of numerous compounds such as DNA and lipid materials, both in living organisms and in food products. From the past, synthetic antioxidants include butylated hydroxytoluene (BHT), butylated hydroxyanisole (BHA), propyl gallate (PG), and tert-butylhydroquinone (TBHQ) have been commercialized and supplemented in some available food products (Shahidi 2000). Even if the synthetic antioxidants have an effective antioxidant ability (Shahidi 2000, Gutteridge and Halliwell 1994, Naczk and Shahidi 2004), they are also reported the adverse effects on human physiological functions (Liu et al. 2016, Winata and Lorenz 1996). Thus, numerous recent studies have been done to explore natural antioxidant compounds and safe, antioxidant peptides (Sila and Bougatef 2016). Antioxidant peptides derived from food origins have been considered as ideally perfect natural antioxidants in the medical and industrial applications (Singh, Vij, and Hati 2014, Yan et al. 2015). They can be safe and potentially reveal a beneficial health effect due to their great activity, simple absorption, and low toxicity (Dei Piu et al. 2014). Fermented seafood products, especially fermented shrimps, are reported that they having high antioxidant activity (Faithong et al. 2010, Peralta et al. 2008, Binsan et al. 2008, Kleekayai, Saetae, et al. 2015, Jung, Rajapakse, and Kim 2005, Park, Je, and Kim 2005). The aqueous extracts of *ka-pi* (MW < 500 Da) showed high 2,2'-azino-bis (3-ethyl-benzthiazoline-6-sulphonic acid ($ABST^+$) radical scavenging activity and this dipeptide was identified, Trp-Pro. $ABST^+$ radical scavenging activity of Trp-Pro was reported as EC_{50} value 17.52 ± 0.46 μM that was closely related to the activity seen with ascorbic acid (EC_{50} = 13.34 ± 0.38 μM) (Kleekayai, Harnedy, et al. 2015). Water-soluble fraction of Thai fermented shrimp samples (*kung-som* and *ka-pi*) exhibited antioxidant activity. *Kung-som* and *ka-pi* had 2,2-diphenyl-1-picryl-hydrazyl-hydrate (DPPH) radical-scavenging activity of 8.58–5.84 and 15.5–4.34 μmol TE/g sample (TE: Trolox equivalents), respectively. The results of $ABTS^+$ radical-scavenging activity of soluble fractions from *kung-som* and *ka-pi* were 67.9-57.6 and 124-376 μmol TE/g sample, respectively. Also, the results of Ferric Reducing Antioxidant Power (FRAP) of soluble fractions from *kung-som* and *ka-pi* were 17.0-24.8 and 48.2-19.5 μmol TE/g sample, respectively (Faithong et al. 2010). Additionally, the maillard reaction products (MRPs) with potential antioxidant activity can be found in the long-period fermentation due to proteins interact with reducing sugars (Peralta et al. 2008). A similar correlation between antioxidant activity and browning intensity was shown in *ka-pi* during the fermentation period. DPPH, ABTS radical scavenging activity as well as FRAP gradually increased as the fermentation time increased. The combination between antioxidant peptides and MRPs may slow down lipid oxidation in fermented shrimp paste (Faithong and Benjakul 2014). Not only antioxidant peptides and MRPs but also coenzyme Q (CoQ; ubiquinone) is lipid-soluble compound, synthesized in human living cells, which participate in the mitochondrial electron-transport chain. The reduced form of CoQ10 (ubiquinol) is a strong antioxidant (Awadallah 2013, Kaikkonen et al. 2002, Chow 2004). Protection against oxidative damage by coenzyme Q has been demonstrated in biological membranes, DNA, lipoproteins, liposomes, low-density, and proteins (Forsmark et al. 1991, Forsmark-Andree and Ernster 1994, Stocker, Bowry, and Frei 1991). The CoQ10 content in the fermented products showed in high level, which was attributed to the contribution of the diverse microflora during the fermentation process .Furthermore, taurine is another biologically active compound founded in salt-fermented shrimp paste, that can modulate the oxidative stress (Peralta et al. 2008). Taurine plays role in beneficial physiological actions such as regulation of mitochondrial protein synthesis; consequently, enhancing electron-transport chain activity and protecting the mitochondria against immoderate superoxide generation)Jong, Azuma, and Schaffer (2012. Carr et al. (1996) reported that the muscle tissue of crustaceans (i.e., shrimp and krill) and mollusks is composed of high levels of taurine. These results indicated that fermented shrimp products (*kung-som* and *ka-pi*) contained compounds which were electron donors and these compounds could convert free radicals to stable metabolites and terminate the radical chain reaction. Therefore, fermented shrimp products could be used as a source of nutrients and natural antioxidants.

6.2. Antihypertensive Activity

Blood pressure is importantly regulated by the renin angiotensin (RAS) and the kallikrein-kinin system (KKS), and influenced by numerous factors (i.e., heart rate, arterial contraction, blood volume, and the nervous system). Hypertension (elevated blood pressure) is the serious medical condition and a high risk factor for myocardial infarction, atherosclerosis, stroke and end-stage renal disease (Chen and Wang 2017). Angiotensin converting enzyme (ACE; EC 3.4.15.1) is a Zn-metallopeptidase that plays an important role in the regulation of peripheral blood pressure in mammals by converting angiotensin-I into angiotensin-II (a vasoconstrictor molecule). For the KKS, Bradykinin transformed into a vaso-inactive peptide by the hydrolysis of ACE. Thus, the inhibition of ACE is a key therapeutic approach to control and reduce hypertension by reduction of the vasoconstric activity of angiotensin-II, and enhancement of the vasodilating activity of bradykinin (Petrillo Jr and Ondetti 1982, Flambard and Johansen 2007, Skeggs, Kahn, and Shumway 1956). Numerous fermented shrimps with Angiotensin converting enzyme (ACE)-inhibitory activity have been reported such as Thai traditional fermented shrimps. The aqueous extracts of two fermented shrimps (*ka-pi*; MW < 500 Da) showed effective ACE inhibition and two dipeptides were identified, Ser-Val and Ile-Phe. These dipeptides revealed ACE inhibitory activity with IC_{50} values 60.68 ± 1.06 μM and 70.03 ± 1.45 μM, respectively (Kleekayai, Harnedy, et al. 2015). From this study, it is possibly indicated that the administration of Thai fermented shrimp products having bioactive peptides with great ACE inhibitory activity may have potential to support consumer health. However, *in vivo* studies are necessarily required to examine and validate these *in vitro* findings.

6.3. Fibrinolytic Activity

Cardiovascular diseases (CVD) are the major cause of the mortality and include myocardial infarction, ischemic heart disease, valvular heart disease, peripheral vascular disease, arrhythmias, hypertension and stroke (Labarthe and Dunbar 2012, Roth et al. 2011). The variety of diseases and health problems are influenced by reactions that maintain a balance between blood anticoagulation and coagulation. Historically, the therapeutic approaches of heart disease and stroke were relied on the use of anticoagulant drugs and anti-platelet drugs to prevent thrombus formation (an excessively inappropriate blood clot), which consists of fibrin and platelets. The accumulation of fibrin in the blood vessels can obstruct the blood circulation and lead to myocardial infarction and other serious cardiovascular diseases. Hence, the appropriately antithrombotic therapeutic approaches should prevent the formation or remove the thrombus. (Stephani et al. 2017, Collen and Lijnen 2004, Kim et al. 1996, Nakajima, Mihara, and Sumi 1993, Wang, Wu, and Liang 2011).

Fibrinolytic enzymes (thrombolytic agents) can lyse preexisting thrombus .These enzymes can also prevent the damage if the clot is early removed after it occurs (Kotb 2012) .Noticeably, fibrinolytic enzymes can be almost found in numerous fermented foods .Especially, fermented shrimp paste, a traditional and popular seasoning in Asian countries, showed strong fibrinolytic activity related to the presence of a novel fibrinolytic enzyme produced during the natural fermentation stage. Moreover, the enzyme was resistant to proteolytic enzymes in *in vitro* assay (Wong and Mine 2004, Mine, Wong, and Jiang 2005). The enzyme, with apparent molecular weight of 18 kDa and N-terminal amino-acid sequence of DPYEEPGPCENLQVA, was specific to fibrin and fibrinogen, without breaking down other blood proteins that are not involved in the blood-clotting cascade. Potentially fibrinolytic activity has been also reported in traditional Korean fermented shrimp (*Jeot-gal*) products (Hwang et al. 2007). Furthermore, *Bacillus* sp. nov. SK006 was isolated from Asian traditional fermented shrimp paste. This strain produced four extracellular fibrinolytic enzymes. One fibrinolytic enzyme was purified to homogeneity with a molecular mass of 43–46 kDa. The enzyme was able to degrade fibrin clots either by forming active plasmin from plasminogen or by direct fibrinolysis. The N-terminal amino acid sequence was found to be AQSVPYEQPHLSQ, that is dissimilar from that of the known fibrinolytic enzymes (Hua et al. 2008). Finally, it is possibly summarized that these finding motivated active searching of numerous fibrinolytic microorganisms in the Asian traditional fermented food and these studies revealed that especially the fermented food is considered as one of the great and safe sources of fibrinolytic enzymes. Nevertheless, further *in vitro* and *in vivo* studies are essentially required to examine and determine characteristics, stabilization, mechanism, metabolism, and safety.

7. Health Risk

From the previous contents, fermented shrimp products are nutritive diet because of its composition, nutrition value and other beneficially bioactive compounds. On the other hand, it can be double-edge sword due to the other compounds associated with health risks such as microorganisms, heavy metals, or other contaminating compounds. Also, biogenic amines can be formed during the fermentation process and high amounts of salt are also usually supplemented to fermentation.

7.1. Biogenic Amines

The biogenic amines synthesis in fermented foods mostly proceeds via the decarboxylation of amino acid carried out by LAB. It is known that the presence of biogenic amines in diverse foods may have adverse effects in humans after consumption. Some of the harmful effects are hypertension, hypotension migraine, and rash (ten Brink et al. 1990). Additionally, it can be effective nitrosamines carcinogenic precursors (Karovičová and Kohajdová 2005). Therefore, the presence of biogenic amines in food has been restricted by different countries and/or organizations. For instance, the Food and Drug Administration (FDA) suggests that the content of histamine, tyramine, and total biogenic amines in seafood products be less than 50 mg/kg (Food and Administration 2002), 100 mg/kg, and 1000 mg/kg respectively (Brillantes and Samosorn 2001). The LAB mainly produces lactic acid which lowers pH and enhances the activity of cathepsins present in muscle. As a result, some amino acids released from native proteins are used as substrate for microbial decarboxylation to synthesize amines. High temperatures also improve the bacterial activity (GÖKOĞLU, YERLIKAYA, and CENGIZ 2004). The tendency to incorporate low salt content and/or carbohydrates contributes to the increase in biogenic amines contents (Köse et al. 2012). The presence of amines in products is depended on various key factors including the fishery season, quality of raw material, and the salting technique (Brillantes, Paknoi, and Totakien 2002). Although the suggested restrictions are not always out of control, the levels of biogenic amines in fermented products are sometimes extremely high. Biogenic amines are thermostabile even in the temperature of an autoclave, which has encouraged the improvement of non-thermal processing to remove or prevent the formation of biogenic amines. The attractive strategy is based on the use of starter cultures with antimicrobial activity and negative decarboxylase activity for controlling the growth of amine-positive bacteria. The study of Hu, Xia, and Liu (2007) found that the inoculation of combined starter cultures of *Lb. casei* subsp. *casei* 1.001, *Lb. plantarum*-15, *Ped. pentosaceus* ATCC 33316, and *Sta. xylosus* 12 can inhibit the growth of contaminant microorganisms which present in the raw materials and also suppress the accumulations of cadaverine, histamine, putrescine, tryptamine, and tyramine. Furthermore, the utilization of halophilic microorganisms in fermented products to degrade biogenic amines has also obtained attention. Biogenic amines can be degraded via amine oxidases (monoamine and diamine oxidases) with the synthesis of aldehyde, ammonia, and hydrogen peroxide (Zaman et al. 2011). It has been discovered that *Sta. xylosus* N°.0538 behaves as a protective culture as it shows an effective ability to degrade histamine and tyramine. This strain can decrease levels of biogenic amines in 16% and may also produce bacteriocin-like compounds with great inhibitory activity against the amine producer *B. licheniformis* (Mah and Hwang 2009). *Sta. carnosus* FS-19 and *B. amyloliquefaciens* FS-05 were able to degrade histamine up to 29.1% and 59.9%, respectively (Zaman et al. 2010). Furthermore, *B. subtilis* FS-12 and *Sta. intermedius* FS-20 were able to degrade cadaverine and putrescine up to 28.9% and 30.4%, respectively. *Natrinema gari* (histamine-degrading bacteria) exhibited great histamine-degrading activity at 40–55°C in the presence of 3.5 to 5 M NaCl and pH 6.5 to 8.0 (Tapingkae et al. 2010). Thus, the use of starter cultures for fermenting seafood products are attractive and important criteria prior use.

7.2. High Amount of Salt Content

Salt is known as the major ingredients and often used in various fermented food production (Lee et al. 2018). Additionally, salt is one of the major substrates contributing to total body fluid balance and especially the primary cause of raised blood pressure (He and MacGregor 2007, Cappuccio 2013). Consequently, a high salt consumption has other adverse effects (He and MacGregor 2007). For example, a direct effect on stroke (Perry and Beevers 1992), demineralized bone (Devine et al. 1995), left ventricular hypertrophy (Kupari, Koskinen, and Virolainen 1994, Schmieder and Messerli 2000), progression of renal disease and albuminuria (Heeg et al. 1989, Cianciaruso et al. 1998, Swift et al. 2005), and gastric cancer (Joossens et al. 1996, Tsugane et al. 2004). Therefore, many health institutions agree that the exceeded salt consumption is an inappropriate behavior and consider salt to be a target for monitoring and prevention (Electrolytes and Water 2005, Graudal and Jurgens 2018). A lessening in salt consumption is one of the most important and impact strategies for improving public health and preventing people dying or suffering (He and MacGregor 2007).

7.3. Heavy Metals

Fermented seafood products may be contaminated from heavy metals (i.e., arsenic (As), cadmium (Cd), lead (Pb), and mercury (Hg)) due to accumulation of these metals in internal organs, mainly in viscera (Bosch et al. 2016). Metal contaminants are normally present in the environments and increasingly amplified by industry and pollution (Erasmus 2007). These toxic metals and metalloids can have adverse effects on the human health. Firstly, Arsenic is mostly detected in meat and fish (Klaassen and Amdur 2013, Castro-González and Méndez-Armenta 2008). Seafood can contain the huge amount of As than other foodstuff (Llobet et al. 2003, Ysart et al. 2000). Acute Symptoms of As exposure in humans include abdominal pain, diarrhea, muscle weakness, skin flushing, vomiting, while chronic As toxicity has led to skin defects and cancer (Schroeder and Darrow 1972, Bosch et al. 2016). Secondly, Cadmium is severely toxic to humans (Erasmus 2007, Ramos-Peralonso 2014). Effects of Cd on human health include cardiovascular function, hypertension, carcinogenic effects, neurological disorders, and skeletal weakness and defects (Schroeder and Darrow 1972, Klaassen and Amdur 2013). Thirdly, Lead is often acquired by the consumption of contaminant food, especially seafood. Excess lead exposure can result in hematological effects, hypertension, neurological problems, renal failure, and cancer (Ebdon et al. 2001, Klaassen and Amdur 2013). Finally, Mercury is the most concern with regards to seafood consumption and human health (Erasmus 2007). Organs of the cardiovascular systems and central nervous are determined to be the critical target organs with regard to methylmercury (MeHg) (Yaktine and Nesheim 2007, Guynup and Safina 2012). Guynup and Safina (2012) reported that low concentration of MeHg can adversely affect the cardiovascular system whereas chronic Hg exposure impacts the liver and the pituitary gland and leads to a compromise of the immune system. Additionally, symptoms of MeHg toxicity in humans include impaired vision and hearing, headaches, paraesthesia, movement difficulties and loss of coordination, fatigue, tremors and ataxia (Grandjean et al. 2010). However, the study of Kawasaki et al. (2008) found that *Staphylococcus* and *Halobacillus* isolated from fermented shrimp and crabs have been used to decrease 30% of Cd content. Thus, efficient overall production and monitoring food safety are important for a prevention of the hazardous compound contamination.

CONCLUSION

In many countries, fermented shrimp products have long been an important food of the human diet. Although these products are consumed in small amount, they can contribute a taste sensation-promoting appetite for eating. Thai traditional fermented shrimp products (*ka-pi, kung-chom,* and *kung-som*) are not only major source of nutritional benefits, especially protein and peptides but also source of beneficial compounds, attractively biological activities (i.e., antioxidant activity, antihypertensive activity, and fibrinolytic activity), and probiotics. Consequently, it can be also concluded that the fermented shrimp products promote consumer health as functional food. However, the inappropriate fermentation and consumption can be the cause of several health risks which lead to some impairment or severe disorders.

REFERENCES

Akolkar, A. V., Durai, D. and Desai, A. J. 2010. "*Halobacterium* sp. SP1 (1) as a starter culture for accelerating fish sauce fermentation." *Journal of applied microbiology* 109, no. 1: 44-53.

Ali, M., Joni, K., Aulanni'am, A., and Yunianta, Y. 2020. "Amino acids, fatty acids and volatile compounds of terasi udang, an Indonesian shrimp paste, during fermentation." *AACL Bioflux* 13:938-950.

Amalia, U. and Agustini, T. W. 2018. "Characterization of Lactic Acid Bacteria (LAB) isolated from Indonesian shrimp paste (terasi)." *E&ES* 116, no. 1: 012049.

Aryanta, W. R. 2000. "Traditional Fermented Foods in Indonesia." *Japanese Journal of Lactic Acid Bacteria* 10:90-102. doi: 10.4109/jslab1997.10.90.

Awadallah, S. 2013. "Protein antioxidants in thalassemia." In *Advances in clinical chemistry*, 85-128. Elsevier.

Aymerich, T., Martin, B., Garriga, M. and Hugas, M. 2003. "Microbial quality and direct PCR identification of lactic acid bacteria and nonpathogenic staphylococci from artisanal low-acid sausages." *Applied and Environmental Microbiology* 69 (8):4583-4594.

Ben Omar, N. and Ampe, F. 2000. "Microbial community dynamics during production of the Mexican fermented maize dough pozol." *Applied and Environmental Microbiology* 66 (9):3664-3673.

Booncharoen, A., Visessanguan, W., Kuncharoen, N., Yiamsombut, S., Santiyanont, P., Mhuantong, W., Charoensri, S., Rojsitthisak, P. and Tanasupawat, S. 2019. "*Lentibacillus lipolyticus* sp. nov., a moderately halophilic bacterium isolated from shrimp paste (Ka-pi)." *International Journal of Systematic and Evolutionary Microbiology* 69 (11):3529-3536.

Bosch, A. C., Bernadette, O., Gunnar, O. S., Sven, E. K. and Louwrens C. H. 2016. "Heavy metals in marine fish meat and consumer health: a review." *Journal of the Science of Food and Agriculture* 96 (1):32-48.

Brillantes, S., Paknoi, S. and Totakien, A. 2002. "Histamine formation in fish sauce production." *Journal of food science* 67 (6):2090-2094.

Brink, B. ten, Damink, C., Joosten, H. M. L. J. and In't Veld, J. H. 1990. "Occurrence and formation of biologically active amines in foods." *International journal of food microbiology* 11 (1):73-84.

Broncano, J. M., Otte, J., Petrón, M.J., Parra, V. and Timón, M. L. 2012. "Isolation and identification of low molecular weight antioxidant compounds from fermented "chorizo" sausages." *Meat science* 90 (2):494-501.

Cai, L., Qingjun, W., Zhijian, D., Shucheng, L., Chaohua, Z. and Jianrong, L. 2017. "Biochemical, nutritional, and sensory quality of the low salt fermented shrimp paste ".*Journal of Aquatic Food Product Technology* 26(6):706-718.

Cappuccio, F. P. 2013. "Cardiovascular and other effects of salt consumption." *Kidney international supplements* 3 (4):312-315.

Carr, W. E. S., Netherton, III., JAMES, C., Richard, A. G. and Charles, D. D. 1996. "Stimulants of feeding behavior in fish: analyses of tissues of diverse marine organisms." *The biological bulletin* 190 (2):149-160.

Castro-González, M. I., and Méndez-Armenta, M. 2008. "Heavy metals: Implications associated to fish consumption." *Environmental toxicology and pharmacology* 26 (3):263-271.

Cha, Y. J., and Cadwallader, K. R. 1995. "Volatile components in salt-fermented fish and shrimp pastes." *Journal of Food Science* 60 (1):19-24.

Chaijan, M. and Panpipat, W. 2012. "Darkening prevention of fermented shrimp paste by pre-soaking whole shrimp with pyrophosphate." *Asian Journal Food and Agro-Industry* 5 (2):163-171.

Chaisri, R. S., Utaipattanacheep, A., Chulakarungka, S., Maungteuk, T. and Maneerat, S. 2005. "Comparative tudies on Chemical Composition, Proteolysis and Chitin Characteristics of Commercial Kung-som and Kung-jom." *Food and Nutrition Program, Department of Home Economics, Faculty of Agriculture, Kasetsart University, Bangkok, Thailand.*

Chaves-López, C., Serio, A., Grande-Tovar, C. D., Cuervo-Mulet, R., Delgado-Ospina, J. and Paparella, A. 2014. "Traditional fermented foods and beverages from a microbiological and nutritional perspective: the Colombian heritage." *Comprehensive Reviews in Food Science and Food Safety* 13 (5):1031-1048.

Chen, L. and Shujun W. 2017. "Preparation of an ACE-inhibitory peptide from *Perinereis aibuhitensis* protein." *Biotechnology & Biotechnological Equipment* 31 (6):1231-1236.

Chow, C. K. 2004. "Dietary coenzyme Q10 and mitochondrial status." In *Methods in enzymology*, 105-112. Elsevier.

Chuea-nongthon, C., Rodtong, S., Yongsawatdigul, J. and Steele, J. L. 2017. "Draft genome sequences of *Tetragenococcus muriaticus* strains 3MR10-3 and PMC-11-5 isolated from Thai fish sauce during natural fermentation." *Genome Announcements* 5 (15).

Chuon, M. R., Shiomoto, M., Koyanagi, T., Sasaki, T., Michihata, T., Chan, S., Mao, S. and Enomoto, T. 2014. "Microbial and chemical properties of Cambodian traditional fermented fish products." *Journal of the Science of Food and Agriculture* 94 (6):1124-1131.

Cianciaruso, B., Bellizzi, V., Minutolo, R., Tavera, A., Capuano, A., Conte, G. and De Nicola, L. 1998. "Salt intake and renal outcome in patients with progressive renal disease." *Mineral and electrolyte metabolism* 24 (4):296-301.

Cocolin, L., Manzano, M., Cantoni, C. and Comi, G. 2001. "Denaturing gradient gel electrophoresis analysis of the 16S rRNA gene V1 region to monitor dynamic changes in the bacterial population during fermentation of Italian sausages." *Applied and Environmental Microbiology* 67 (11):5113-5121.

Collen, D. and Lijnen, H. R. 2004. "Tissue-type plasminogen activator: a historical perspective and personal account." *Journal of Thrombosis and Haemostasis* 2 (4):541-546.

Dangkhaw, N., Maneerat, S. and Sumpavapol, P. 2012. "Characterization of lactic acid bacteria isolated from Kung-Som, a traditional fermented shrimp, in respect of their probiotic properties." *International Conference on Nutrition and Food Sciences* (39).

Daroonpunt, R., Saeng-in, P. and Tanasupawat, S. 2019. "Identification and lipolytic activity of *Bacillus* and *Staphylococcus* strains from shrimp paste (Ka-pi)." *Journal of Applied Pharmaceutical Science* 9 (04):024-029.

Daroonpunt, R., Tanasupawat, S., Kudo, T., Ohkuma, M. and Itoh, T. 2016. "*Virgibacillus kapii* sp. nov., isolated from Thai shrimp paste (Ka-pi)." *International journal of systematic and evolutionary microbiology* 66 (4):1832-1837.

Daroonpunt, R., Uchino, M., Tsujii, Y., Kazami, M., Oka, D. and Tanasupawat, S. 2016. "Chemical and physical properties of Thai traditional shrimp paste (*Ka-pi*)." *Journal Applied Pharmaceutical Science* 6 (05):058-062.

Daroonpunt, R., Yiamsombut, S., Sitdhipol, J. and Tanasupawat, S. 2019. "*Bacillus salacetis* sp. nov., a slightly halophilic bacterium from Thai shrimp paste (Ka-pi)." *International journal of systematic and evolutionary microbiology* 69 (4):1162-1168.

Dei Piu, L., Tassoni, A., Serrazanetti, D. I., Ferri, M., Babini, E., Tagliazucchi, D. and Gianotti, A. 2014. "Exploitation of starch industry liquid by-product to produce bioactive peptides from rice hydrolyzed proteins." *Food chemistry* 155:199-206.

Devine, A., Criddle, R. A., Dick, I. M., Kerr, D. A. and Prince, R. L. 1995. "A longitudinal study of the effect of sodium and calcium intakes on regional bone density in postmenopausal women." *The American journal of clinical nutrition* 62 (4):740-745.

Diana, M., Quílez, J. and Rafecas, M. 2014. "Gamma-aminobutyric acid as a bioactive compound in foods: a review." *Journal of Functional Foods* 10:407-420.

Ebdon, L., Pitts, L., Cornelis, R., Crews, H., Quevauviller, P. and Donard, O. F. X. 2001. *Trace element speciation for environment, food and health*: Royal Society of Chemistry.

Erasmus, C. P. 2007. "*The concentration of ten metals in the tissues of shark species squalus megalops and mustelus mustelus (chondrichthyes) occuring along the southerneastern coast of South Africa.*"

Ercolini, D. 2004. "PCR-DGGE fingerprinting: novel strategies for detection of microbes in food." *Journal of microbiological methods* 56 (3):297-314.

Faithong, N. and Benjakul, S. 2014. "Changes in antioxidant activities and physicochemical properties of Kapi, a fermented shrimp paste, during fermentation." *Journal of Food science and Technology* 51 (10):2463-2471.

Faithong, N., Benjakul, S., Phatcharat, S. and Binsan, W. 2010. "Chemical composition and antioxidative activity of Thai traditional fermented shrimp and krill products." *Food Chemistry* 119 (1):133-140.

Fan, Y., Xue, Y., Li, Z., Hou, H. and Xue, C. 2017. "Analyzing the flavor compounds in Chinese traditional fermented shrimp pastes by HS-SPME-GC/MS and electronic nose." *Journal of Ocean University of China* 16 (2):311-318.

Flambard, B., and Johansen, E. 2007. "Developing a functional dairy product: from research on *Lactobacillus helveticus* to industrial application of Cardi-04™ in novel antihypertensive drinking yoghurts." In *Functional dairy products*, 506-520. Elsevier.

Food, and Drug Administration (FDA). 2002. "FDA's evaluation of the seafood HACCP program for fiscal years 2000/2001." *Report available at http://www. cfsan. fda. gov/~ comm/seaeval2. html.*

Forsmark, P., Åberg, F., Norling, B., Nordenbrand, K., Dallner, G. and Ernster, L. 1991. "Inhibition of lipid peroxidation by ubiquinol in submitochondrial particles in the absence of vitamin E." *FEBS letters* 285 (1):39-43.

Forsmark-Andree, P. and Ernster, L. 1994. "Evidence for a protective effect of endogenous ubiquinol against oxidative damage to mitochondrial protein and DNA during lipid peroxidation." *Molecular aspects of medicine* 15(1):s73-s81.

Gökoğlu, N., Yerlikaya, P. and Cengiz, E. 2004. "Changes in biogenic amine contents and sensory quality of sardine (*Sardina pilchardus*) stored at 4C and 20C." *Journal of Food Quality* 27 (3):221-231.

Grandjean, P., Satoh, H., Murata, K. and Eto, K. 2010. "Adverse effects of methylmercury: environmental health research implications." *Environmental health perspectives* 118 (8):1137-1145.

Graudal, N. and Jürgens, G. 2018. "Conflicting evidence on health effects associated with salt reduction calls for a redesign of the salt dietary guidelines." *Progress in Cardiovascular Diseases* 61 (1):20-26.

Guan, L., Cho, K. H. and Lee, J. H. 2011. "Analysis of the cultivable bacterial community in jeotgal, a Korean salted and fermented seafood, and identification of its dominant bacteria." *Food microbiology* 28 (1):101-113.

Gutteridge, J. and Halliwell, B. 1994. *Antioxidants in nutrition, health, and disease*: Oxford University Press.

Guynup, S. and Safina, B. C. 2012. "Mercury: Sources in the environment, health effects, and politics." *Blue Ocean Institute* 1:7-54.

Hajeb, P. and Jinap, S. 2012. "Fermented shrimp products as source of umami in Southeast Asia." *Journal of Nutrition & Food Sciences* 10:006.

He, F. J. and MacGregor, G. A. 2007. "Dietary salt, high blood pressure and other harmful effects on health." In *Reducing salt in foods*, 18-54. Elsevier.

Heeg, J. E., De Jong, P. E., Van Der Hem, G. K. and De Zeeuw, D. 1989. "Efficacy and variability of the antiproteinuric effect of ACE inhibition by lisinopril." *Kidney international* 36 (2):272-279.

Hill, C., Guarner, F., Reid, G., Gibson, G., Daniel, J. M. R., Pot, B., Morelli, L., Canani, R. B., Flint, H. J. and Salminen, S. 2014. "Expert consensus document: The International Scientific Association for Probiotics and Prebiotics consensus statement on the scope and appropriate use of the term probiotic." *Nature reviews Gastroenterology & hepatology* 11 (8):506.

Hinrichsen, L. L., Montel, M. C. and Talon, R. 1994. "Proteolytic and lipolytic activities of *Micrococcus roseus* (65), *Halomonas elongata* (16) and *Vibrio* sp. (168) isolated from Danish bacon curing brines." *International journal of food microbiology* 22 (2-3):115-126.

Horrocks, L. A., and Yeo, Y. K. 1999. "Health benefits of docosa-hexaenoic acid (DHA)." *Pharmacological research* 40 (3):211-225.

Hu, Y., Xia, W. and Liu, X. 2007. "Changes in biogenic amines in fermented silver carp sausages inoculated with mixed starter cultures." *Food Chemistry* 104 (1):188-195.

Hua, Y., Jiang, B., Mine, Y. and Mu, W. 2008. "Purification and characterization of a novel fibrinolytic enzyme from *Bacillus* sp. nov. SK006 isolated from an Asian traditional fermented shrimp paste." *Journal of agricultural and food chemistry* 56 (4):1451-1457.

Hwang, K., Choi, K., Kim, M., Park, C. and Cha, J. 2007. "Purification and characterization of a new fibrinolytic enzyme of *Bacillus licheniformis* KJ-31, isolated from Korean traditional Jeot-gal." *Journal of microbiology and biotechnology* 17 (9):1469.

Hwanhlem, N., Watthanasakphuban, N., Riebroy, S., Benjakul, S., H-Kittikun, A. and Maneerat, S. 2010. "Probiotic lactic acid bacteria from Kung-Som: isolation, screening, inhibition of pathogenic bacteria." *International journal of food science & technology* 45 (3):594-601.

Hyne, R. V., Sánchez-Bayo, F., Bryan, A. D., Johnston, E. L. and Mann, R. M. 2009. "Fatty acid composition of the estuarine amphipod, Melita plumulosa (Zeidler): link between diet and fecundity." *Environmental Toxicology and Chemistry: An International Journal* 28 (1):123-132.

Ichimura, T., Hu, J., Aita, D. Q. and Maruyama, S. 2003. "Angiotensin I-converting enzyme inhibitory activity and insulin secretion stimulative activity of fermented fish sauce." *Journal of bioscience and bioengineering* 96 (5):496-499.

Institute of Medicine (US). Panel on Dietary Reference Intakes for, and Water. 2005. *DRI, dietary reference intakes for water, potassium, sodium, chloride, and sulfate*: National Academy Press.

Irianto, H. E. 2017. "Enzymes in fermented fish." In *Advances in Food and Nutrition Research*, 199-216. Elsevier.

Ishige, N. 1993. "Cultural aspects of fermented fish products in Asia." *Fish fermentation technology*.

Ishikawa, M., Tanasupawat, S., Nakajima, K., Kanamori, H., Ishizaki, S., Kodama, K., Okamoto-Kainuma, A., Koizumi, Y., Yamamoto, Y. and Yamasato, K. 2009. "*Alkalibacterium thalassium* sp. nov., *Alkalibacterium pelagium* sp. nov., *Alkalibacterium putridalgicola* sp. nov. and *Alkalibacterium kapii* sp. nov., slightly halophilic and alkaliphilic marine lactic acid bacteria isolated from marine organisms and salted foods collected in Japan and Thailand." *International journal of systematic and evolutionary microbiology* 59 (5):1215-1226.

Jinap, S., Ilya-Nur, A. R., Tang, S. C., Hajeb, P., Shahrim, K. and Khairunnisak, M. 2010. "Sensory attributes of dishes containing shrimp paste with different concentrations of glutamate and 5′-nucleotides." *Appetite* 55 (2):238-244.

JIRCAS. 2011a. "*Fermented shrimp (Kung-chom).*" Last Modified 15/04/2020, accessed 30/07/2020. https://www.jircas.affrc.go.jp/DB/DB11/fish8803.html?id=7.

JIRCAS. 2011b. "*Fermented shrimp (Kung-som).*" 2011, accessed 30/07/2020. https://www.jircas.affrc.go.jp/DB/DB11/fish0055.html?id=11.

JIRCAS. 2011c. "*Fermented shrimp paste (Kapi).*" Last Modified 15/04/2020, accessed 30/07/2020. https://www.jircas.affrc.go.jp/DB/DB11/fish190a.html?id=17.

Jong, C. J., Azuma, J. and Schaffer, S. 2012. "Mechanism underlying the antioxidant activity of taurine: prevention of mitochondrial oxidant production." *Amino acids* 42 (6):2223-2232.

Joossens, J. V., Hill, M. J., Elliott, P., Stamler, R., Stamler, J., Lesaffre, E., Dyer, A., Nichols, R., Kesteloot, H., European Cancer Prevention and the Intersalt Cooperative Research Group. 1996. "Dietary salt, nitrate and stomach cancer mortality in 24 countries." *International journal of epidemiology* 25 (3):494-504.

Jung, W. K., Rajapakse, N. and Kim, S. K. 2005. "Antioxidative activity of a low molecular weight peptide derived from the sauce of fermented blue mussel, *Mytilus edulis*." *European Food Research and Technology* 220 (5-6):535-539.

Kaewklom, S., Lumlert, S., Kraikul, W. and Aunpad, R. 2013. "Control of *Listeria monocytogenes* on sliced bologna sausage using a novel bacteriocin, amysin, produced by *Bacillus amyloliquefaciens* isolated from Thai shrimp paste (Kapi)." *Food Control* 32 (2):552-557.

Kaikkonen, J., Tuomainen, T. P., Nyyssönen, K. and Salonen, J. T. 2002. "Coenzyme Q10: absorption, antioxidative properties, determinants, and plasma levels." *Free radical research* 36 (4):389-397.

Karovičová, J. and Kohajdová, Z. 2005. "Biogenic amines in food." *Chemical papers* 59 (1):70-79.

Kawasaki, K., Matsuoka, T., Satomi, M., Ando, M., Tukamasa, Y. and Kawasaki, S. 2008. "Reduction of cadmium in fermented squid gut sauce using cadmium-absorbing bacteria isolated from food sources." *Journal of food, agriculture & environment* 6(1): 45-49.

Kechagia, M., Basoulis, D., Konstantopoulou, S., Dimitriadi, D., Gyftopoulou, K., Skarmoutsou, N. and Fakiri E. M. 2013. "Health benefits of probiotics: a review." *ISRN nutrition* 2013.

Kim, W., Choi, K., Kim, Y., Park, H., Choi, J., Lee, Y., Oh, H., Kwon, I. and Lee, S. 1996. "Purification and characterization of a fibrinolytic enzyme produced from *Bacillus* sp. strain CK 11-4 screened from Chungkook-Jang." *Applied and Environmental Microbiology* 62 (7):2482-2488.

Kim, Y. B., Choi, Y. S., Ku, S. K., Jang, D. J., binti Ibrahim, H. H. and Moon, K. B. 2014. "Comparison of quality characteristics between belacan from Brunei Darussalam and Korean shrimp paste." *Journal of Ethnic Foods* 1 (1):19-23.

Klaassen, C. D. and Amdur, M. O. 2013. *Casarett and Doull's toxicology: the basic science of poisons*. Vol. 1236: McGraw-Hill New York.

Kleekayai, T., Harnedy, P. A., O'Keeffe, M. B., Poyarkov, A. A., Cunhaneves, A., Suntornsuk, W. and FitzGerald, R. J. 2015. "Extraction of antioxidant and ACE inhibitory peptides from Thai traditional fermented shrimp pastes." *Food chemistry* 176:441-447.

Kleekayai, T., Saetae, D., Wattanachaiyingyong, O., Tachibana, S., Yasuda, M. and Suntornsuk, W. 2015. "Characterization and in vitro biological activities of Thai traditional fermented shrimp pastes." *Journal of Food Science and Technology* 52 (3):1839-1848.

Kobayashi, T., Kajiwara, M., Wahyuni, M., Kitakado, T., Hamada-Sato, N. Imada, C. and Watanabe, E. 2003. "Isolation and characterization of halophilic lactic acid bacteria isolated from "terasi" shrimp paste: a traditional fermented seafood product in Indonesia." *The Journal of General and Applied Microbiology* 49 (5):279-286.

Köse, S., Koral, S., Tufan, B., Pompe, M., Scavniçar, A. and Koçar, D. 2012. "Biogenic amine contents of commercially processed traditional fish products originating from European countries and Turkey." *European Food Research and Technology* 235 (4):669-683.

Kotb, E. 2012. "Fibrinolytic bacterial enzymes with thrombolytic activity." In *Fibrinolytic Bacterial Enzymes with Thrombolytic Activity*, 1-74. Springer.

Kumaunang, M., Sanchart, C., Suyotha, W. and Maneerat, S. 2019. "*Virgibacillus halodenitrificans* MSK-10P, a Potential Protease-producing Starter Culture for Fermented Shrimp Paste (kapi) Production." *Journal of Aquatic Food Product Technology* 28 (8):877-890.

Kupari, M., Koskinen, P. and Virolainen J. 1994. "Correlates of left ventricular mass in a population sample aged 36 to 37 years. Focus on lifestyle and salt intake." *Circulation* 89 (3):1041-1050.

Labarthe, D. R. and Dunbar, S. B. 2012. "Global cardiovascular health promotion and disease prevention: 2011 and beyond." *Circulation* 125 (21):2667-2676.

Lee, K. W., Shim, J. M., Kim, D. W., Yao, Z., Kim, J. A., Kim, H. J. and Kim, J. H. 2018. "Effects of different types of salts on the growth of lactic acid bacteria and yeasts during kimchi fermentation." *Food science and biotechnology* 27 (2):489-498.

Lee, S. H., Jung, J. Y. and Jeon, C. O. 2015. "Bacterial community dynamics and metabolite changes in myeolchi-aekjeot, a Korean traditional fermented fish sauce, during fermentation." *International journal of food microbiology* 203:15-22.

Leong, Q. L., Ab Karim, S., Selamat, J., Mohd Adzahan, N., Karim, R. and Rosita, J. 2009. "Perceptions and acceptance of 'belacan'in Malaysian dishes." *International Food Research Journal* 16:539-546.

Liu, R., Xing, L., Fu, Q., Zhou, G. H. and Zhang, W. G. 2016. "A review of antioxidant peptides derived from meat muscle and by-products." *Antioxidants* 5 (3):32.

Llobet, J. M., Falco, G., Casas, C., Teixido, A. and Domingo, J. L. 2003. "Concentrations of arsenic, cadmium, mercury, and lead in common foods and estimated daily intake by children, adolescents, adults, and seniors of Catalonia, Spain." *Journal of Agricultural and Food Chemistry* 51 (3):838-842.

Lobo, V., Patil, A., Phatak, A. and Chandra, N. 2010. "Free radicals, antioxidants and functional foods: Impact on human health." *Pharmacognosy reviews* 4 (8):118.

Lv, X., Li, Y., Cui, T., Sun, M., Bai, F., Li, X., Li, J. and Yi, S. 2020. "Bacterial community succession and volatile compound changes during fermentation of shrimp paste from Chinese Jinzhou region." *LWT -Food Science and Technology* 122:108998.

Mah, J. H. and Hwang H. J. 2009. "Inhibition of biogenic amine formation in a salted and fermented anchovy by *Staphylococcus xylosus* as a protective culture." *Food Control* 20 (9):796-801.

Martínez-Álvarez, O., López-Caballero, M. E., Gómez-Guillén, M. C. and Montero, P. 2017. "Fermented seafood products and health." In *Fermented foods in health and disease prevention*, 177-202. Elsevier.

Mine, Y., Wong, A. H. K. and Jiang, B. 2005. "Fibrinolytic enzymes in Asian traditional fermented foods." *Food Research International* 38 (3):243-250.

Montaño, N., Gavino, G. and Gavino, V. C. 2001. "Polyunsaturated fatty acid contents of some traditional fish and shrimp paste condiments of the Philippines." *Food Chemistry* 75 (2):155-158.

Murwani, R., Subagio, S., Trianto, A. and Ambariyanto. 2015. "Isolation and identification of thermophilic and mesophylic proteolytic bacteria from shrimp paste "Terasi"." *AIP Conference Proceedings.*

Naczk, M. and Shahidi, F. 2004. "Extraction and analysis of phenolics in food." *Journal of chromatography A* 1054 (1-2):95-111.

Najafian, L. and Babji, A. S. 2015. "Isolation, purification and identification of three novel antioxidative peptides from patin (*Pangasius sutchi*) myofibrillar protein hydrolysates." *LWT-Food Science and Technology* 60 (1):452-461.

Nakajima, N., Mihara, H. and Sumi, H. 1993. "Characterization of potent fibrinolytic enzymes in earthworm, *Lumbricus rubellus*." *Bioscience, biotechnology, and biochemistry* 57 (10):1726-1730.

Namwong, S., Tanasupawat, S., Lee, K. C. and Lee, J. S. 2009. "*Oceanobacillus kapialis* sp. nov., from fermented shrimp paste in Thailand." *International journal of systematic and evolutionary microbiology* 59 (9):2254-2259.

Nantida, D. 2018. "Fermented Shrimps (Kung-Som) with Lactic Acid Bacteria to Cure a Disease." *Dusit Thani College Journal* 11 (special):359-376.

Ohashi, Y. and Ushida, K. 2009. "Health-beneficial effects of probiotics: Its mode of action." *Animal Science Journal* 80 (4):361-371.

Padilah, B., Bahruddin, S., Fazilah, A., Ahmad, M. and Gulam, R. R. A. 2018. "Biogenic amines analysis in shrimp pastes belacan obtained from the Northern States of Peninsular Malaysia." *International Food Research Journal* 25 (5):1893-1899.

Pakawadee, I. 2016. "Effect of salt on fermentation and sensory characteristics of Kung-Jom." *Journal of King Mongkut's University of Technology North Bangkok* 26 (1):105-112.

Pakdeeto, A., Tanasupawat, S., Thawai, C., Moonmangmee, S., Kudo, T. and Itoh, T. 2007a. "*Lentibacillus kapialis* sp. nov., from fermented shrimp paste in Thailand." *International journal of systematic and evolutionary microbiology* 57 (2):364-369.

Pakdeeto, A., Tanasupawat, S., Thawai, C., Moonmangmee, S., Kudo, T. and Itoh, T. 2007b. "*Salinicoccus siamensis* sp. nov., isolated from fermented shrimp paste in Thailand." *International journal of systematic and evolutionary microbiology* 57 (9):2004-2008.

PARK, P. J., JE, J. Y. and KIM, S. K. 2005. "Amino acid changes in the Korean traditional fermentation process for blue mussel, Mytilus edulis." *Journal of food biochemistry* 29(1): 108-116.

Peralta, E. M., Hatate, H., Kawabe, D., Kuwahara, R., Wakamatsu, S., Yuki, T. and Murata, H. 2008. "Improving antioxidant activity and nutritional components of Philippine salt-fermented shrimp paste through prolonged fermentation." *Food Chemistry* 111 (1):72-77.

Perry, I. J. and Beevers, D. G. 1992. "Salt intake and stroke: a possible direct effect." *Journal of human hypertension* 6 (1):23-25.

Petrillo Jr, E. W. and Ondetti, M. A. 1982. "Angiotensin- converting enzyme inhibitors: medicinal chemistry and biological actions." *Medicinal research reviews* 2 (1):1-41.

Pham-Huy, L. A., He, H. and Pham-Huy, C. 2008. "Free radicals, antioxidants in disease and health." *International journal of biomedical science: IJBS* 4 (2):89.

Phewpan, A., Phuwaprisirisan, P., Takahashi, H., Ohshima, C., Lopetcharat, K., Techaruvichit, P. and Keeratipibul, S. 2020. "Microbial diversity during processing of Thai traditional fermented shrimp paste, determined by next generation sequencing." *LWT-Food Science and Technology* 122:108989.

Phithakpol, B. 1993. "Fish fermentation technology in Thailand." In *Fish fermentation technology*, eds. Lee, C.H., Steinkraus, K.H. & Alan Reilly, P.J., 155-166. Tokyo: United Nations University Press.

Phithakpol, B. and Kasetsat, M. 1995. "*The traditional fermented foods of Thailand*: *Institute of food research and product development*." Kasetsart University.

Phoottosavako, M., Keeratipibul, S., Techo, S. and Tanasupawat, S. 2015. "Identification and characterization of lipolytic bacteria from Thai fermented foods." *Malaysian Journal of Microbiology* 11 (3):231-239.

Pihlanto-Leppälä, A., Rokka, T. and Korhonen, H. 1998. "Angiotensin I converting enzyme inhibitory peptides derived from bovine milk proteins." *International Dairy Journal* 8 (4):325-331.

Pilapil, A. R., Neyrinck, E., Deloof, D., Bekaert, K., Robbens, J. and Raes, K. 2016. "Chemical quality assessment of traditional salt-fermented shrimp paste from Northern Mindanao, Philippines." *Journal of the Science of Food and Agriculture* 96 (3):933-938.

Pongsetkul, J. 2018. "*Properties and Changes during Fermentation of Salted Shrimp Paste as Affected by Quality of Row Material and Selected Halophilic Bacteria*." Prince of Songkla University.

Pongsetkul, J., Benjakul, S., Sampavapol, P., Osako, K. and Faithong, N. 2014. "Chemical composition and physical properties of salted shrimp paste (Kapi) produced in Thailand." *International Aquatic Research* 6 (3):155-166.

Pongsetkul, J., Benjakul, S., Sumpavapol, P., Vongkamjan, K. and Osako, K. 2018. "*Bacillus subtilis* K-C3 isolated from Thai salted shrimp paste (*Kapi*): Its extracellular enzymes and use as a starter culture in Kapi production." *Journal of food biochemistry* 42 (6):e12649.

Pongsetkul, J., Benjakul, S., Vongkamjan, K., Sumpavapol, P. and Osako, K. 2017. "Microbiological and chemical changes of shrimp *Acetes vulgaris* during Kapi production." *Journal of food science and technology* 54 (11):3473-3482.

Prachasitthisak, Y., Eamsiri, J., Sajjabut, S. and Rojekittikhun, W. 2008. "Improving the hygiene of fermented shrimp (kung chom) using gamma irradiation."*Journal of Tropical Medicine and Parasitology* 32(1):9-16.

Prapasuwannakul, N. and Suwannahong, K. 2015. "Chemical composition and antioxidant activity of Klongkone shrimp paste." *Procedia-Social and Behavioral Sciences* 197:1095-1100.

Ramos-Peralonso, M. J. 2014. “European Food Safety Authority (EFSA) Scientific Opinion on Dietary Reference Values for iodine.” *EFSA J* 12:1-57.

Rattagool, P. 1985. “Fermented fish products of South East Asia.” *Tropical Science* 25:61-73.

Rattanachaikunsopon, P. and Phumkhachorn, P. 2008. "Incidence of nisin Z production in *Lactococcus lactis* subsp. *lactis* TFF 221 isolated from Thai fermented foods." *Journal of food protection* 71 (10):2024-2029.

Roth, G. A., Fihn, S. D., Mokdad, A. H., Aekplakorn, W., Hasegawa, T. and Lim, S. S. 2011. "High total serum cholesterol, medication coverage and therapeutic control: an analysis of national health examination survey data from eight countries." *Bulletin of the World Health Organization* 89:92-101.

Saelao, S., Keasuwan, S., Benjakul, S. and Maneerat, S. 2016. “Safety evaluation and bacterial community of kung-som using PCR-DGGE technique.” *Songklanakarin Journal of Science & Technology* 38 (4):413-419.

Saelao, S., Maneerat, S., Kaewsuwan, S., Rabesona, H., Choiset, Y., Haertlé, T. and Chobert, J. M. 2017. "Inhibition of *Staphylococcus aureus in vitro* by bacteriocinogenic *Lactococcus lactis* KTH0-1S isolated from Thai fermented shrimp (Kung-som) and safety evaluation." *Archives of microbiology* 199 (4):551-562.

Salampessy, J., Kailasapathy, K. and Thapa, N. 2010. “Fermented fish products.” *Fermented foods and beverages of the world*:289-307.

Sanchart, C., Benjakul, S., Rattanaporn, O., Haltrich, D. and Maneerat, S. 2015. “Efficiency of the V3 region of 16S rDNA and the rpoB gene for bacterial community detection in Thai traditional fermented shrimp (Kung-Som) using PCR-DGGE techniques.” *Songklanakarin Journal of Science & Technology* 37 (3):291-297.

Sanchart, C., Rattanaporn, O., Haltrich, D., Phukpattaranont, P. and Maneerat, S. 2017. "*Lactobacillus futsaii* CS3, a new GABA-producing strain isolated from Thai fermented shrimp (Kung-Som)." *Indian journal of microbiology* 57 (2):211-217.

Sangjindavong, M., Limpakdee, J., Wilaipun, P. and Vareevanich, D. 2006. "Utilization of Antibacterial Substances from Lactic Acid Bacteria for Extension Shelf-life of Chilled Crab Meat." *Agriculture and Natural Resources* 40 (3):702-707.

Schmieder, R. E. and Messerli, F. H. 2000. "Hypertension and the heart." *Journal of human hypertension* 14 (10):597-604.

Schroeder, H. A. and Darrow, D. K. 1972. "Relation of trace metals to human health." *Envtl. Aff.* 2:222.

Shahidi, F. 2000. "Antioxidants in food and food antioxidants." *Food/nahrung* 44 (3):158-163.

Shahidi, F. and Zhong, Y. 2008. "Bioactive peptides." *Journal of AOAC international* 91 (4):914-931.

Sikorski, Z. E., Kołakowska, A. and Pan, B. S. 1990. "The nutritive composition of the major groups of marine food organisms." *The nutritive composition of the major groups of marine food organisms* 29-54.

Sila, A. and Bougatef, A. 2016. "Antioxidant peptides from marine by-products: Isolation, identification and application in food systems. A review." *Journal of Functional Foods* 21:10-26.

Singh, B. P., Vij, S. and Hati, S. 2014. "Functional significance of bioactive peptides derived from soybean." *Peptides* 54:171-179.

Skeggs Jr, L. T., Kahn, J. R. and Shumway, N. P. 1956. "The preparation and function of the hypertensin-converting enzyme." *The Journal of experimental medicine* 103 (3):295.

Stephani, L., Tjandrawinata, R. R., Afifah, D. N., Lim, Y., Ismaya, W. T. and Suhartono, M. T. 2017. "Food origin fibrinolytic enzyme with multiple actions." *HAYATI Journal of Biosciences* 24 (3):124-130.

Stocker, R., Bowry, V. W. and Frei, B. 1991. "Ubiquinol-10 protects human low density lipoprotein more efficiently against lipid peroxidation than does alpha-tocopherol." *Proceedings of the National Academy of Sciences* 88 (5):1646-1650.

Sudaryati, Y. and Rahmansyah, M. 2014. "*Bacillus subtilis* A9 producing lipase isolated from shrimp paste." *Global Journal of Biology, Agriculture & Health Science* 3:205-210.

Supapun, B., and Samosorn, W. 2001. "Determination of histamine in fish sauce from Thailand using a solid phase extraction and high-performance liquid chromatography." *Fisheries science* 67 (6):1163-1168.

Surono, I. S., and Hosono, A. 1994a. "Chemical and aerobic bacterial composition of "Terasi", a traditional fermented product from Indonesia." *Food Hygiene and Safety Science (Shokuhin Eiseigaku Zasshi)* 35(3):299-304.

Surono, I. S., and Hosono, A. 1994b. "Microflora and their enzyme profile in "*Terasi*" starter." *Bioscience, biotechnology, and biochemistry* 58 (6):1167-1169.

Swift, P. A., Markandu, N. D., Sagnella, G. A., He, F. J. and MacGregor, G. A. 2005. "Modest salt reduction reduces blood pressure and urine protein excretion in black hypertensives: a randomized control trial." *Hypertension* 46 (2):308-312.

Tanasupawat, S., Hashimoto, Y., Ezaki, T., Kozaki, M. and Komagata, K. 1992. "*Staphylococcus piscifermentans* sp. nov., from fermented fish in Thailand." *International Journal of Systematic and Evolutionary Microbiology* 42 (4):577-581.

Tanasupawat, Somboon. 2009. "Thai lactic acid bacteria: diversity and applications." *Srinakharinwirot Science Journal* 25 (1).

Tapingkae, W., Tanasupawat, S., Parkin, K. L., Benjakul, S. and Visessanguan, W. 2010. "Degradation of histamine by extremely halophilic archaea isolated from high salt-fermented fishery products." *Enzyme and Microbial Technology* 46 (2):92-99.

TCPS324. 2004. "*Thai Community Product Standard*."

Thongruck, K., Saelao, S., Sumpavapol, P., Benjakul, S. and Maneerat, S. 2017. "Monitoring of changes in lactic acid bacteria during production of Thai traditional fermented shrimp (*Kung-Som*) by culturing method and PCR-DGGE technique." *Songklanakarin Journal of Science and Technology* 39 (1):41-47.

Todd, J. J. 2014. "Lactate: valuable for physical performance and maintenance of brain function during exercise." *Bioscience Horizons: The International Journal of Student Research* 7.

Tsugane, S., Sasazuki, S., Kobayashi, M. and Sasaki, S. 2004. "Salt and salted food intake and subsequent risk of gastric cancer among middle-aged Japanese men and women." *British Journal of Cancer* 90 (1):128-134.

Udomsil, N., Rodtong, S., Choi, Y. J., Hua, Y. and Yongsawatdigul, J. 2011. "Use of *Tetragenococcus halophilus* as a starter culture for flavor improvement in fish sauce fermentation." *Journal of agricultural and food chemistry* 59 (15):8401-8408.

Visessanguan, W., Benjakul, S., Smitinont, T., Kittikun, C., Thepkasikul, P. and Panya, A. 2006. "Changes in microbiological, biochemical and physico-chemical properties of Nham inoculated with different inoculum levels of *Lactobacillus curvatus*." *LWT-Food Science and Technology* 39 (7):814-826.

Wang, S. L., Wu, Y. Y. and Liang, T. W. 2011. "Purification and biochemical characterization of a nattokinase by conversion of shrimp shell with *Bacillus subtilis* TKU007." *New biotechnology* 28 (2):196-202.

Wanwisa, B., Benjakul S., Visessanguan, W., Roytrakul, S., Tanaka, M. and Hideki Kishimura. 2008. "Antioxidative activity of Mungoong, an extract paste, from the cephalothorax of white shrimp (*Litopenaeus vannamei*)." *Food Chemistry* 106 (1):185-193.

Winata, A. and Lorenz, K. 1996. "Antioxidant potential of 5-n-pentadecylresorcinol." *Journal of food processing and preservation* 20 (5):417-429.

Wittanala, S., Rakariyatham, N. and Deming, R. L. 2011. "Volatile compounds of vegetarian soybean kapi, a fermented Thai food condiment." *African Journal of Biotechnology* 10 (5):821-830.

Wong, A. H. K. and Mine, Y. 2004. "Novel fibrinolytic enzyme in fermented shrimp paste, a traditional Asian fermented seasoning." *Journal of Agricultural and Food Chemistry* 52 (4):980-986.

Yaktine, A. L. and Nesheim, M. C. 2007. *Seafood choices: balancing benefits and risks*: National Academies Press.

Yan, Q. J., Huang, L. H., Sun, Q., Jiang, Z. Q. and Wu, X. 2015. 2015. "Isolation, identification and synthesis of four novel antioxidant peptides from rice residue protein hydrolyzed by multiple proteases." *Food chemistry* 179:290-295.

Yoshida, Y. 1998. "Umami taste and traditional seasonings." *Food Reviews International* 14 (2-3):213-246.

Ysart, G., Miller, P., Croasdale, M., Crews, H., Robb, P., Baxter, M., De L'Argy, C. and Harrison, N. 2000. "1997 UK Total Diet Study dietary exposures to aluminium, arsenic, cadmium, chromium, copper, lead, mercury, nickel, selenium, tin and zinc." *Food Additives & Contaminants* 17 (9):775-786.

Zaman, M. Z., Bakar, F. A., Jinap, S. and Bakar, J. 2011. "Novel starter cultures to inhibit biogenic amines accumulation during fish sauce fermentation." *International journal of food microbiology* 145 (1):84-91.

ZaMaN, M. Z., Bakar, F. A., SelaMat, J. and Bakar, J. 2010. "Occurrence of biogenic amines and amines degrading bacteria in fish sauce." *Czech Journal of Food Sciences* 28 (5):440-449.

Zhang, T. T., Xu, J., Wang, Y. M. and Xue, C.H. 2019. "Health benefits of dietary marine DHA/EPA-enriched glycerophospholipids." *Progress in Lipid Research* 75: 100997.

INDEX

A

B

C

D

E

F

R

S

T

U

V

W

Z